Marx, Joyce
Abbott Hardy Montaigne Chesterton Austen
Defoe Melville Machiavelli Cooper Emerson Hugo
Stoker Carroll Haggard Molière Eliot Grimm
Wilde Christie Maupassant Byron Schiller
Garnett Fitzgerald Engels
Goethe Einstein Hawthorne Kafka
Cotton Dostoyevsky Smith Hall
Baum Kipling Doyle Willis
Dumas Henry Nietzsche
Leslie Flaubert Turgenev Balzac
Stockton Vatsyayana Crane
Burroughs Verne
Curtis Tocqueville Gogol Vinci
Homer Widger Tolstoy Whitman Busch
Darwin Thoreau
Potter Freud Twain Scott
Zola Lawrence Plato Harte
Kant Jowett Stevenson Dickens Hesse
Andersen Cervantes Burton
London Descartes Voltaire
Poe Aristotle Wells
Hale James Hastings Cooke
Bunner Shakespeare Irving
Richter Chambers Cid
Doré da Shaw Benedict Alcott
Dante Chekhov Wodehouse Pushkin
Swift Newton

tredition

tredition was established in 2006 by Sandra Latusseck and Soenke Schulz. Based in Hamburg, Germany, tredition offers publishing solutions to authors and publishing houses, combined with worldwide distribution of printed and digital book content. tredition is uniquely positioned to enable authors and publishing houses to create books on their own terms and without conventional manufacturing risks.

For more information please visit: www.tredition.com

TREDITION CLASSICS

This book is part of the TREDITION CLASSICS series. The creators of this series are united by passion for literature and driven by the intention of making all public domain books available in printed format again - worldwide. Most TREDITION CLASSICS titles have been out of print and off the bookstore shelves for decades. At tredition we believe that a great book never goes out of style and that its value is eternal. Several mostly non-profit literature projects provide content to tredition. To support their good work, tredition donates a portion of the proceeds from each sold copy. As a reader of a TREDITION CLASSICS book, you support our mission to save many of the amazing works of world literature from oblivion. See all available books at www.tredition.com.

 Project Gutenberg

The content for this book has been graciously provided by Project Gutenberg. Project Gutenberg is a non-profit organization founded by Michael Hart in 1971 at the University of Illinois. The mission of Project Gutenberg is simple: To encourage the creation and distribution of eBooks. Project Gutenberg is the first and largest collection of public domain eBooks.

Science and Morals and Other Essays

Bertram Coghill Alan, Sir Windle

Imprint

This book is part of TREDITION CLASSICS

Author: Bertram Coghill Alan, Sir Windle
Cover design: Buchgut, Berlin – Germany

Publisher: tredition GmbH, Hamburg - Germany
ISBN: 978-3-8472-1633-9

www.tredition.com
www.tredition.de

Copyright:
The content of this book is sourced from the public domain.

The intention of the TREDITION CLASSICS series is to make world literature in the public domain available in printed format. Literary enthusiasts and organizations, such as Project Gutenberg, worldwide have scanned and digitally edited the original texts. tredition has subsequently formatted and redesigned the content into a modern reading layout. Therefore, we cannot guarantee the exact reproduction of the original format of a particular historic edition. Please also note that no modifications have been made to the spelling, therefore it may differ from the orthography used today.

SCIENCE AND MORALS

SCIENCE AND MORALS

AND OTHER ESSAYS

BY

SIR BERTRAM C. A. WINDLE
M.A., M.D., Sc.D., LL.D., F.R.S., F.S.A., K.S.G.
OF ST. MICHAEL'S COLLEGE, TORONTO, ONT.

LONDON
BURNS & OATES, LTD
28 ORCHARD STREET, W
1919

TO

JOHN ROBERT and MARY O'CONNELL

A TOKEN OF SINCERE FRIENDSHIP

Listarkin
September 1919

THESE Essays have all in one form or another appeared elsewhere; and I have to thank the Editors of the *Dublin Review*, *Catholic World*, *America*, and *Studies* respectively for kind permission to reproduce them. Some of them appear as they were published, others have been almost rewritten.

B. C. A. W.

[Pg 1]

SCIENCE AND MORALS

SCIENCE AND MORALS

§ 1. THE GOSPEL OF SCIENCE

In the days before the war the Annual Address delivered by the President of the British Association was wont to excite at least a mild interest in the breasts of the reading public. It was a kind of Encyclical from the reigning pontiff of science, and since that potentate changed every year there was some uncertainty as to his subject and its treatment, and there was this further piquant attraction, wanting in other and better-known Encyclicals, that the address of one year might not merely contradict but might even exhibit a lofty contempt for that or for those which had immediately preceded it.

During the three years immediately preceding the war we had excellent examples of all these things. In the first of them we were treated to a somewhat belated utterance in opposition to Vitalism. Its arguments were mostly based upon what even to the tyro in chemistry seemed to be rather shaky foundations. Such indeed they proved to be, since the deductions drawn from [Pg 2] the behaviour of colloids and from Leduc's pretty toys were promptly disclaimed by leading chemists in the course of the few days after the delivery of the address.

Further, the President for the year 1914 in his address (Melbourne, p. 18) [1] told us that the problem of the origin of life, which, let us remind ourselves, in the 1912 address was on the point of solution, "still stands outside the range of scientific investigation," and that when the spontaneous formation of formaldehyde is talked of as a first step in that direction he is reminded of nothing so much as of Harry Lauder, in the character of a schoolboy, "pulling his treasures from his pocket—'That's a wassher—for makkin motor-cars!'" Nineteen hundred and twelve pinned its faith on matter and nothing else; Nineteen hundred and thirteen assured us that "occurrences now regarded as occult can be examined and reduced to order by the methods of science carefully and persistently applied."

[2] Further, the examination of those facts had convinced the deliverer of the address "that memory and affection are not limited to that association with matter by which alone they can manifest themselves here and now, and that personality persists beyond bodily death." Nineteen hundred and fourteen proclaimed telepathy a "harmless toy," which, with necromancy, has taken the place [Pg 3] of "eschatology and the inculcation of a ferocious moral code." And yet it is on telepathy, if we are to believe the daily papers, that Sir Oliver Lodge largely relies for his proofs. Here, at any rate, is a pleasing diversity of opinion which fully bears out what was said at the beginning of this paper. It is, however, with the third address, or rather pair of addresses, that we are concerned; for the meeting of 1914, not only was the first to be held at the Antipodes, but also the first to be honoured with two addresses — one in Melbourne, the other in Sydney.

Their deliverer is a very distinguished and a very independent man of Science. It was he who insisted, at a time when the domination of a very rigid form of Darwinism was much stronger than it is to-day, that the picture of Nature as seen by us is a Discontinuous picture, though Discontinuity does not exist in the environment. And it was he who asked whether the Discontinuity might not be in the living thing itself, and prefixed to the monumental work [3] in which he discussed this question the significant text from the Bible: "All flesh is not the same flesh; but there is one kind of flesh of men, another flesh of beasts, another of fishes, and another of birds." Nearer to our own times, he was one of a small body of men of science who almost synchronously disinterred the forgotten works of Abbot Mendel, and proclaimed them to the world, as containing discoveries of the first value. He was thus always something of a "Herald of Revolt," and maintains [Pg 4] that character in these addresses. "We go to Darwin for his incomparable collection of facts. We would fain emulate his scholarship, his width and his power of exposition, but to us he speaks no more with philosophical authority. We read his scheme of evolution as we would those of Lucretius or Lamarck, delighting in their simplicity and their courage" (M., p. 9). "Naturally, we turn aside from generalities. It is no time to discuss the origin of the Mollusca or of Dicotyledons, while we are not even sure how it came to pass that *Primula obconica* has in twenty-five years produced its abundant new forms almost under our eyes" (*ib., ib.*). And so on. To take one other example: there is nothing which was more insisted upon by Darwinians than the fact that all the various races of domestic fowl known to us came from *Gallus bankiva*, the jungle-fowl of India; in fact I think I have seen that form enthroned amongst its supposed descendants in more than one museum. "So we are taught; but try to reconstruct the steps in their evolution and you realise your hopeless ignorance" (M., p. 11). If we cannot construct a "tree" for fowls, how absurd to adventure into the deeper recesses of Phylogeny. If all that Professor Bateson says is true, is not Driesch right when he speaks of "the phantasy christened Phylogeny"? [4]

The addresses, however, were not solely concerned with throwing contempt upon views which were yesterday of great respectability, and which even to-day are as gospel to many. They [Pg 5] devoted themselves chiefly to the consideration of the question of heredity, viewed, as might be expected, from the Mendelian standpoint.

Now, at this point it may be said that there are at least two things which we should like to know about heredity — the vehicle and the laws. It is clear that we might know something, perhaps even a good deal, about one of these without knowing anything about the other.

Such in fact is the case; for we know, it may fairly be said, nothing about the vehicle. There are two very widely distinct opinions on this point. There is the mnemic theory, recently brought before us by the republication of Butler's most interesting and suggestive work with its translations of Hering's original paper and Von Hartmann's discourse and its very illuminating introduction by Professor Hartog. [5]

And there is the continuity theory which teaches that in some way or another the characteristics of the parents and other ancestors are physical parts of the germ. An attempt to explain this was made by Darwin in his theory of Pangenesis. Others have essayed what Yves Delage calls "micromeristic" interpretations. As to all of these it may be said that when they are reduced to figures the explanation becomes of so complex a character as utterly to break down. We shall see that Professor Bateson adopts a third very nebulous explanation. But as regards the laws of heredity there is something else to be said; for here we really do know something, and [Pg 6] that something we owe in large measure to the innumerable experiments which have been made on Mendelian lines since the rediscovery of the methods first adopted by the celebrated Abbot of Brünn. It is no intention of the writer of this paper to describe the Mendelian theory, [6] which is well known, at least to all biological readers, though one or two points in connection with it may yet have to be touched upon.

The point of cardinal importance in connection with Mendelism is that it does reveal a law capable of being numerically stated, and apparently applicable to a large number of isolated factors in living things. Indeed it was this attention to isolated factors which was the first and essential part of Mendel's method. For example, others had been content to look at the pea as a whole. Mendel applied his analytic method to such things as the colour of the pea, the smooth or wrinkled character of the skin which covered it, its dwarfness or height, and so on.

Now, the behaviour of these isolated factors seems to throw a light even upon the vehicle of heredity. We often talk of "blood" and "mixing of blood," as if blood had anything to do with the question, when really the Biblical expression "the seed of Abraham" is much more to the point. For it is in the seed that these factors must be, whether they be mnemic or [Pg 7] physical. Professor Bateson (M., p. 5) thinks it obvious that they are transmitted by the spermatozoon and the ovum; but it seems to him "unlikely that they are in any simple or literal sense material particles." And he goes on to say, and this, I think, is one of his most important statements: "I suspect rather that their properties depend on some phenomenon of arrangement."

Now, if there be a law behind the phenomena made clear to us by Mendelian experiments (as Mendelians are never tired of asserting), then it becomes in no way impertinent to ask how that law came into existence, and who formulated it. Darwinism, according to Driesch, [7] "explained how by throwing stones one could build houses of a typical style." In other words, it "claimed to show how something purposively constructed could arise by absolute chance; at any rate this holds of Darwinism as codified in the seventies and eighties." Of course the Blind Chance doctrine breaks down utterly when it comes to be applied to selected cases, and nothing more definitely disposes of it than the very definite law which emerges as the result of the Mendelian experiments. That is obvious to the prophets of Mendelism; but, whilst they admit this, they will have nothing to say to the lawgiver. That is the "rankest metaphysics," as Dr. Johnstone puts it, [8] or "mysticism," as others prefer to call it. And yet nothing is more clear than the logical [Pg 8] sequence that, if you have a law, someone must have made it, and if you look upon something as "a phenomenon of arrangement," someone must have arranged it. But for reasons not obvious nor confessed, there is an objection to make any such admission. Perhaps it is the taint of the monism of the latter half of the last century which still persists.

At any rate, as I have elsewhere pointed out, there is a most curious passage in another paper by the same author in which he says: "With the experimental proof that variation consists largely in the unpacking and repacking of an original complexity, it is not so certain as we might like to think that the order of these events is not pre-determined." The writer hastens to denounce the horrid heresy on the brink of which he finds himself hesitating, by adding that he sees "no ground whatever for holding such a view," though "in the light of modern research it scarcely looks so absurdly improbable as before." [9] It is curious that the writer in question does not seem to have been in any way influenced by the eliminative argument so potent in connection with the discussion on Vitalism. We ask for an explanation of the occurrences—say of regeneration. We find that no physical explanation in the least meets the needs of the case, and we are consequently obliged to look for it in something differing from the operations of chemistry and physics. Of this argument Dr. John [Pg 9] stone [10] says: "It is almost impossible to overestimate the appeal which it makes to the investigator."

Now, this matter of "arrangement" or of "pre-determination," when put forward as an explanation, even tentatively, necessitates a step further. That step might possibly be in the direction of pantheism, though, according to Driesch, [11] pantheism is the doctrine "that reality is a something which makes itself (*'dieu se fait,'* in the words of Bergson), whilst theism would be any theory according to which the manifoldness of material reality is predetermined in an immaterial way." And he concludes "that those who regard the thesis of the theory of order as necessary for everything that is or can be, must accept theism, and are not allowed to speak of '*dieu qui se fait*.'" It is difficult to see how anyone who has studied the rigid order exhibited by experiments on Mendelian lines can resist the logic of this argument unless indeed he takes a place on Plate's platform, which admits that a law entails a lawgiver, but declares that of the Lawgiver of Natural Laws we can know nothing. [12]

There is a further point in connection with Mendelian theories which is worth noting in this connection. It would appear that no new factor is ever brought into being, that is, no *addition* is ever made by variation. According to this [Pg 10] theory the things which appear to be added — a new colour or a new scent — were there all the time. They were "stopped down" or inhibited by some other factor, which, when eliminated, allows them to come into play, and thus to become obvious to the observer from whom they had been hidden. Thus, Professor Bateson (M., p. 17) has confidence "that the artistic gifts of mankind will prove to be due, not to something added to the make-up of an ordinary man, but to the absence of factors which in the normal person inhibit the development of these gifts. They are almost beyond doubt to be looked upon as *releases* of powers normally suppressed. The instrument is there, but it is 'stopped down.'"

That all sorts of things may exist in a very small compass no doubt is true. Professor Bateson reminds us that Shakespeare was once "a speck of protoplasm not so big as a small pin's head." The difficulty — insuperable on ordinary monistic lines — is how all these things got into the germ if no additions ever take place. It was so difficult to account, for example, for artistic appreciation on the part of man or for gifts of an artistic character that Huxley was fain to describe them as gratuitous; but on this showing all characters are gratuitous in the sense that they are not acquired. We may reasonably inquire not merely how all these characters and factors got themselves "arranged" or "packed," but where they came from, and how they came to be in the germ at all, matters on which we receive no information in these addresses. No [Pg 11] doubt the author of the addresses would say that it was no part of his business to explain this matter; that he took this system of Nature as a going system and did his best to explain it as such and without attempting, perhaps even without desiring, to explain how it got a-going. If that be the case, and if ignorance on this head must be his confession, it is a little difficult to understand the confidence with which he sets himself to discuss the "extraordinary and far-reaching changes in public opinion [which] are coming to pass." We shall find these, as we pass them in review, to be extraordinary enough, though not very new.

In the first place, "genetic research will make it possible for a nation to elect by what sort of beings it will be represented not very many generations hence, much as a farmer can decide whether his byres shall be full of shorthorns or Herefords. It will be very surprising indeed if some nation does not make trial of this new power. They may make awful mistakes, but I think they will try" (S., p. 8). It is curious how the war, which had just commenced when these addresses were being delivered, has absolutely disposed, or ought to have disposed, of some of the prophecies of the President. Nothing, at any rate, seems more certain than that one result of this most disastrous struggle will be an urgent demand by all the States engaged in it for at least as many male children as the mothers of each country can supply, without special regard to their other characters, breedable or not breedable. We are [Pg 12] even told that Germany is resorting to expedients which cannot be justified on Christian principles to fill her depleted homes. Whether this be true or not the fact remains that nothing is now more to be desired by all the combatant nations than what we call in Ireland "long families." But even if there had been no war, there is one other factor which makes it quite certain that no country ever will try, or if it ventures to try, will ever succeed in any such experiment, and that factor, forgotten by philosophers of this kind, is human nature. Mr. Frankfort Moore years ago wrote a pleasant story, called "The Marriage Lease," in which doctrinaire legislation of a somewhat similar kind was described, and its inevitable failure most amusingly depicted. The war disposes of another of the President's maxims (S., p. 10), that the decline in the birth-rate of a country is nothing to be grieved about, and that "the slightest acquaintance with biology" shows that the "inference may be wholly wrong," which asserts that "a nation in which population is not rapidly increasing must be in a decline" (S., p. 10). Human nature was neglected in the first-mentioned case, and here it is the turn of history to pass into the shade, history which, *pace* the President, has really a good deal more bearing upon a question of this kind than the "school-boy natural history" which he thinks capable of settling it. Thus we advance from breeding to Malthusianism. It is perhaps not wonderful that our next step should be the quiet, and of course painless, extinction of the unfit.

[Pg 13]

> "Thou shalt not kill, but needs't not strive
> Officiously to keep alive."

Thus wrote Clough; but our author, it appears, would go further than this. "The preservation of an infant so gravely diseased that it can never be happy or come to any good is something very like wanton cruelty. In private life few men defend such interference" (S. 10). And so such unfortunates should be got rid of, and will be "as soon as scientific knowledge becomes common property"—when "views more reasonable, and, I may add, more humane are likely to prevail." Lest we should be depressed by this massacre of the innocents, we are told that "man is just beginning to know himself for what he is—a rather long-lived animal, with great powers of enjoyment if he does not deliberately forgo them" (S., p. 9). In the past, poor fool that he has been, he has not availed himself of his opportunities: "Hitherto superstition and mythical ideas of sin have predominantly controlled these powers." Let us, however, take heart: "Mysticism will not die out; for those strange fancies knowledge is no cure; but their forms may change, and mysticism as a force for the suppression of joy is happily losing its hold on the modern world" (*ib., ib.*). Let us eat and drink—and, it may be added, sin—for to-morrow we die. Such is the new gospel of science, an old enough gospel, tried and found wanting years before its latest prophet arose to proclaim it to the world. Surely no more ridiculous utterance ever was made; for its author evidently [Pg 14] did not pause to consider that the sins which make life pleasant to some (for example, Thuggery) are apt to have quite another aspect to those through whose victimisation the pleasure is obtained. There is also here such a thing as the conscience, which has to be taken into account. Even the biological hedonist must originally possess such a thing and, it may be supposed, must deal with it as he would with the gravely diseased children, and as something which would "predominantly control his powers of enjoyment."

Seriously, it may be doubted if a more pagan code of morals has ever been laid down, and this in the Encyclical of Science for the year, a code bad enough to make poor Mendel turn in his grave could he—good, honest man—be aware of it, and imagine that he was in any way responsible for it, which, by the way, is in no way the case.

§ 2. SCIENCE AS A RULE OF LIFE

Saint or sinner, some rule of life we must have, even if we are wholly unconscious of the fact. A spiritual director will help us to map out a course of action which will assist us to shake off some little of the dust of this dusty world; and a doctor will lay down for us a dietary which will help us to elude, for a time at least, the insidious onsets of the gout. Even if we take no formal steps, spiritual or corporeal, some rule of life we must achieve for ourselves. We must, for example, [Pg 15] make up our minds whether we are to open our ears and our purse to tales of misery, or are to join ourselves with those whose rule of life it is to keep that which they have for themselves. What is true of each of us is none the less true of each and every race—even more true; for each race must make up its mind definitely as to which rule it will follow. And at the moment there is still doubt and indecision in this matter.

"The moral problem that confronts Europe to-day is: What sort of righteousness are we, individually and collectively, to pursue? Is the new righteousness to be realised in a return to the old brutality? Shall the last values be as the first? Must ethical process conform to natural process as exemplified by the life of any animal that secures dominancy at the expense of the weaker members of its kind?" [13] Such are the questions raised by a man of science occupying the

Presidential Chair of an important society and speaking to that society as its President.

As to the Christian ideals little need be said, since we know very well what they are, and know this most especially, that practically all of them are in direct opposition to what we may call the ideals of Nature, and exercise all their influence in frustrating such laws as that of Natural Selection. "Nature's Insurgent Son," as Sir Ray Lankester calls him, [14] is at constant war with Nature, and when we come to consider the matter carefully, [Pg 16] in that respect most fully differentiates himself from all other living things, none of which make any attempt to control the forces of Nature for their own advantage. "Nature's inexorable discipline of death to those who do not rise to her standard — survival and parentage for those alone who do — has been from the earliest times more and more definitely resisted by the will of man. If we may for the purpose of analysis, as it were, extract man from the rest of Nature, of which he is truly a product and a part, then we may say that man is Nature's rebel. Where Nature says 'Die!' man says 'I will live.'" [15]

To this it may be added that, under the influence of Christianity, man goes a step further and says: "I will endeavour that as many others as may be shall live, and live happy, healthy lives, and shall not untimely die." The law of Natural Selection could not be met by more direct opposition. I have said that this is under the influence of Christianity, yet the impulse seems to be older than that, to be part of that moral law which excited Kant's admiration, which he coupled with the sight of the starry heavens, an impulse, we can scarcely doubt, implanted in the heart of man by God Himself. It is a remarkable fact that in many — some would say most — of the less civilised races of mankind we find these social virtues, which some would have us believe are degenerate features foisted on to the race by an enervating superstition.

Dr. Marett has carefully examined into this [Pg 17] matter, and his conclusions are of the greatest interest. [16]

"My own theory about the peasant, as I know him, and about people of lowly culture in general so far as I have learnt to know about them, is that the ethics of amity belong to their natural and normal mood, whereas the ethics of enmity, being but 'as the shadow of a passing fear,' are relatively accidental. Thus to the thesis that human charity is a by-product, I retort squarely with the counter-thesis that human hatred is a by-product. The brute that lurks in our common human nature will break bounds sometimes; but I believe that whenever man, be he savage or civilised, is at home to himself, his pleasure and pride is to play the good neighbour. It may be urged by way of objection that I overestimate the amenities, whether economic or ethical, of the primitive state; that a hard life is bound to produce a hard man. I am afraid that the psychological necessity of the alleged correlation is by no means evident to me. Surely the hard-working individual can find plenty of scope for his energies without needing, let us say, to beat his wife. Nor are the hard-working peoples of the earth especially notorious for their inhumanity. Thus the Eskimo, whose life is one long fight against the cold, has the warmest of hearts. Mr. Stefanson says of his newly discovered 'Blonde Eskimo,' a people still living in the stone age: 'They are the equals of the best of our own [Pg 18] race in good breeding, kindness, and the substantial virtues.' [17] Or again, heat instead of cold may drive man to the utmost limit of his natural affections. In the deserts of Central Australia, where the native is ever threatened by a scarcity of food, his constant preoccupation is not how to prey on his companions. Rather he unites with them in guilds and brotherhoods, so that they may feast together in the spirit, sustaining themselves with the common hope and mutual suggestion of better luck to come. But there is no need to go so far afield for one's proofs. I appeal to those who have made it their business to be intimate with the folk of our own countryside. Is it not the fact that unselfishness in regard to the sharing of the necessaries of life is characteristic of those who find them most difficult to come by? The poor are by no means the least 'rich towards God.' At any rate, if poverty sometimes hardens, wealth, especially sudden wealth, can harden too, causing arrogance, boastfulness, and the bullying temper. 'A proud look, a lying tongue, and the shedding of innocent blood'—these go together."

On the whole, then, we may perhaps conclude that the natural bias of mankind is towards kindness to his neighbour, however much the brute in him may sometimes impel him to uncharitable words or actions. And certainly this natural bias is intensified and made into a binding law by the teachings of Christ. But there is the other [Pg 19] point of view set forward in the philosophy of Nietzsche—if indeed such writings are worthy of the name philosophy. "The world is for the superman. Dominancy within the human kind must be secured at all costs. As for the old values, they are all wrong. Christian humility is a slavish virtue; so is Christian charity. Such values have become 'denaturalised.' They are the by-product of certain primitive activities, which were intended by Nature to subserve strictly biological ends, but have somehow escaped from Nature's control and run riot on their own account."

The prophets of this group of ideals, or some such group of ideals, have no hesitation in telling us how they would direct the affairs of humanity if they were entrusted with their conduct. It will not be without interest to consider their plans and to endeavour to form some sort of an idea of what kind of place the world would be if they had their way. We can then form our own opinion as to whether a world conducted on such lines would be in any way a tolerable place for human existence.

First of all we may dwell briefly on Natural Selection as a rule of life, since it has been put forward as such by quite a number of persons. Never, let it at once be said, by the great and gentle-hearted originator of that theory, who during his life had to protest as to the ignorant and exaggerated ideas which were expressed about it and who, were he now alive, would certainly be shocked at the teachings which are supposed to [Pg 20] follow from his theory and the dire results which they have produced. [18]

In the first place such a doctrine leads directly to the conclusion that war, instead of being the curse and disaster which all reasonable people, not to say all Christians, feel it to be, is, as Bernhardi puts it, "a biological necessity, a regulative element in the life of mankind that cannot be dispensed with." It is "the basis of all healthy development." "Struggle is not merely the destructive but the life-giving principle. The law of the strong holds good everywhere. Those forms survive which are able to secure for themselves the most favourable conditions. The weaker succumb." Humanity has had at times evidences of the results of this teaching which are not, one may fairly say, of a kind to commend themselves to any person possessed of a moderately kindly, not to say of a Christian, disposition. Fortunately, or unfortunately, we have the opportunity of studying the experiment in actual operation in a race which, of course in entire ignorance of the fact, is actually putting into practice the teachings of Natural Selection, though it must be admitted that the practice has not been successful, nor does it look like being successful, in raising that race above the very lowest rung of the ladder of civilisation. Captain Whiffen [19] has given a very complete and a very [Pg 21] interesting account of the peoples whom he met with during his wanderings in the regions indicated by the title of his book. And he tells us that "the survival of the most fit is the very real and the very stern rule of life in the Amazonian forests. From birth to death it rules the Indians' life and philosophy. To help to preserve the unfit would often be to prejudice the chances of the fit. There are no arm-chair sentimentalists to oppose this very practical consideration. The Indian judges it by his standard of common sense: why live a life that has ceased to be worth living when there is no bugbear of a hell to make one cling to the most miserable of existences rather than risk greater misery?" Let us now see the kind of life which the author, freed himself no doubt from "the bugbear of hell," considers eminently sensible — the kind of life of which only an "arm-chair sentimentalist" would disapprove; a kind of life, it may be added, which will appear to most ordinarily minded people as being one of selfishness raised to its highest power.

To begin with the earliest event in life. If a child, on its appearance in the world, appears to be in any way defective, its mother quietly kills it and deposits its body in the forest. If the mother dies in childbirth the child, unless someone takes pity on it and adopts it, is killed by the father, who, it may be presumed, is indisposed to take the trouble, perhaps indeed incapable of doing so, of rearing the motherless babe. That the child, in any case, immediately after birth, is [Pg 22] plunged into cold water, is not perhaps a conscious method of eliminating the weak, though it must operate in that direction. At a later period of life should any disease believed to be infectious break out in a tribe, "those attacked by it are immediately left, even by their closest relatives, the house is abandoned, and possibly even burnt. Such derelict houses are no uncommon sight in the forest, grimly desolate mementoes of possible tragedies." When a person becomes insane, he is first of all exorcised by the medicine man, and if that fails is put to death by poison by the same functionary. The sick are dealt with on similar lines, unless there is or seems to be a probability of speedy recovery. "Cases of chronic illness meet with no sympathy from the Indians. A man who cannot hunt or fight is regarded as useless, he is merely a burden on the community." Under these circumstances he is either left at home untended or hunted out into the bush to die, or his end is accelerated by the medicine man. The same fate awaits the aged, unless they seem to be of value to the tribe on account of their wisdom and experience.

All these things placed together give us a perfect picture of life under Natural Selection, and having studied it we may fairly ask whether such a rule of life is one under which any one of us would like to live. In every respect it is the antipodes of the Christian rule of life, and of that rule of life which civilised countries, whether in fact Christian or not, have derived from Christianity and still practise. The non-Christian [Pg 23] rule of the Indians is one under which might is right and no real individual liberty exists, all personal rights being sacrificed to the supposed needs and benefit of the community.

So much from the point of view of Natural Selection, but it would appear that those who have given up that factor as of anything but a very minor value, if even that, have also their rule of life founded on their interpretation of Nature. Thus Professor Bateson, the great exponent of Mendel's doctrines, who has told us in his Presidential Address to the British Association that we must think much less highly of Natural Selection than some would have us do, has, as has been set forth in the previous section of this essay, his opinion as to the rule of life which we should follow.

Professor Conklyn, an American enthusiast for extreme eugenistic views, has also set down in print his ideas as to the lines on which our lives are to be run under a scientific domination, and these are to be dealt with in another article. [20] His scheme entails a forcible visit, not, it may be supposed, to the Altar, but to the Registry Office, for all persons held to be fit to perpetuate the race, and forcible restraint, whether by imprisonment or by sterilisation, for all others.

The first thing which all these essays towards a scientific conduct of life reveal is a total want of perspective, for they proceed on the hypothesis—which no doubt their authors would defend—that this world and its concerns are everything, and [Pg 24] that the intellectual and physical improvement of the human race by any measures, however harsh, is the "one thing needful." But beyond this the persons who hold such views seem to have entirely overlooked the fact that their proposed State would be one conducted on principles of the bitterest and most galling slavery imaginable by the mind of man, a form of slavery that never could persist if for a moment it be conceded that it could ever come into operation. The fact is that the whole thing is ludicrous when looked at from the point of view of common sense, but how few take the trouble to contemplate these schemes as they would be in operation! Were they thus to contemplate them, they would see that, apart altogether from any religious considerations, they are wholly impossible, even from a purely political point of view. That such ideas are intolerable to Catholic minds, indeed to any Christian mind, goes without saying.

Driesch (*Science and Philosophy of the Organism*, vol. ii., p. 358) has pointed out very clearly that "the mechanical theory of life is incompatible with morality," and that it is impossible to feel "morally" towards other individuals if one knows that they are machines and nothing more. Again, Professor Henslow (in *Present Day Rationalism Critically Examined*, p. 253) very pertinently asks those who discard all religious considerations and claim to rely for guidance on the lessons of Nature, "If you have no taste for virtue, why be virtuous at all, so long as you do not violate the laws of the land?"

[Pg 25]

Yet, in the face of these surely obvious facts, we find persons making such absurd claims as that made in a recent book by Rignano, an Italian writer (*Essays in Scientific Synthesis*, 1917). It is not often that one meets a book so full of philosophical fallacies as this. "We are certain of one fact," he says, "that the only organ actually brought into play to fight immorality is the organ of the collective conscience and not the religious organ." I suppose no more ludicrously inaccurate remark ever was set down in print; for, to begin with, the "collective conscience," whatever that may be, does not exist in Nature, *teste* the farmyard and the fowl-run; and again, whatever force is connoted by those words must have been set agoing—by what? By Nature? Oh, most emphatically No! Nature has no law against immorality; there is no Categorical Imperative in Nature commanding us to be chaste or kindly or considerate or even just. We must go elsewhere if we are to look for teaching in the virtues. That is the fact that we must keep clearly before our minds when endeavouring to estimate at their proper value the nostrums of writers such as those with whose works we have been dealing.

FOOTNOTES:

[Pg 26]

[1] Two addresses were delivered in 1914—one in Melbourne, the other in Sydney. These will be referred to in this article as M. & S.

[2] Sir Oliver Lodge: *Continuity*, p. 90.

[3] *Materials for the Study of Variation*, London, 1894.

[4] *The History and Theory of Vitalism*, p. 140.

[5] *Unconscious Memory*. Fifield. 1910.

[6] Those who desire further information may be referred to *A Century of Scientific Thought*, by the present writer. Burns & Oates.

[7] *Op. cit.*, pp. 137-8.

[8] *The Philosophy of Biology*, p. 64.

[9] In an article in the volume *Darwin and Modern Science*, p. 100.

[10] *Op. cit.*, p. 319.

[11] *Op. cit.*, pp. 238-9.

[12] See the discussion on this subject in Wasmann's *The Problem of Evolution*.

[13] R. R. Marett, Presidential Address to Folk-Lore Society, 1915. *Folk-Lore*, vol. xxvii., pp. 1-14.

[14] *The Kingdom of Man*. London: Constable & Co. 1907.

[15] Lankester, *op. cit.*, p. 26.

[16] *Op. cit.*, pp. 21-27.

[17] *My Life with the Eskimo* (1913), p. 188.

[18] For a discussion of this question, see *Bernhardi and Creation*, by Sir James Crichton-Browne, F.R.S. Glasgow: James Maclehose & Sons. 1916.

[19] *The Northwest Amazons*. London: Constable & Co. 1915.

[20] *Science and the War*, p. 120.

II. THEOPHOBIA AND NEMESIS

§ 1. THEOPHOBIA: ITS CAUSE

Initium sapientiæ timor Domini; no doubt, but such fear is only the beginning, and is not the kind of fear—which also exists—a fear which engenders an actual revulsion against the idea of God.

It is to this kind of fear which the eminent Jesuit writer Wasmann alludes when he says that "in many scientific circles there is an absolute *Theophobia*, a dread of the Creator. I can only regret this," he continues, "because I believe that it is due chiefly to a defective knowledge of Christian philosophy and theology."

That he is entirely right as to the existence of this feeling there can be no doubt; no one can read at all widely in scientific literature without becoming aware of it. Contrary to all the tenets of science there is even a bias against any such idea as that of a Creator, though science is supposed to confront all problems without bias of any kind. I need not cite instances of this feeling; I have dealt with it elsewhere. We may take it for granted, and proceed to look for an explanation for the phenomenon. Wasmann attributes it to ignorance, and he is, I feel sure, right; but let us [Pg 27] examine the matter a little more closely. Why should persons—even if ignorant—have the bias which some obviously present against the idea of a God? Why should they wish to think that there is no such Being, no future existence, nothing higher than Nature? Some persons maintain that precedent to a denial of God there must be a moral failure. That I am sure is quite wrong. I should be far from saying that in some materialists there is not a considerable weakening of moral fibre, or perhaps it would be better put, a distortion of moral vision, as evidenced by many of the statements and proposals of eugenists, for example, and by the political nostrums of some who wrest science to a purpose for which it was not intended. This no doubt is true, but it is not quite the argument with which I am now dealing, and that argument, if it implies moral failure in the persons concerned, has little if any genuine foundation in fact. Mr. Devas, in that very remarkable book, *The Key to the World's Progress*, gives us the useful phrase "post-Christians." These people are really pagans living in the Christian era, retaining many of the excellent qualities which they owe neither to Nature nor to paganism, but to the inheritance—perhaps involuntary and unrecognised—of the influences of Christianity. Many of these people are kind, benevolent, scrupulously moral. They have not learned to be such from Nature, for Nature teaches no such lessons. Nor have they learnt them from paganism, for these are not pagan virtues. They are an inheritance from Chris [Pg 28] tianity. Those, therefore, who build arguments as to the needlessness of religion on the foundation that persons without any belief in God do exhibit all the moral virtues, build on sand. At any rate the answer to the question which we are discussing is not to be found in this direction.

Others again will perhaps maintain the thesis that fashion has a great deal to do with this. It is not fashionable to believe in God, or at least it was not. It was highly fashionable to call oneself an agnostic; perhaps it is not quite so much the vogue now as it was. No doubt there is something in this, though not very much. It is much easier to go with the tide than against it, and there are scientific tides as truly as there are tides in the fashion of dress. There was a Weismann tide, now nearly at dead water; there was an antivitalistic tide, now ebbing fast. When these were in full flow it was a hazardous thing for a young man who had to make his own way in the scientific world to swim against either or both of them. Fashions change, and fashion is not so set against the idea of a God as it was. The materialistic tide is "going out," and we shall see that there is some truth in the view which holds that the incoming tide is largely that of occultism, a thing disliked and despised—and indeed with some reason—by the materialistic school even more than it dislikes and despises theistic opinions.

Fashion, however, is not in any way a complete answer to the question we are proposing to ourselves, nor is the unquestionable fact that scientific [Pg 29] men have a strong objection to putting their trust in anything which cannot be subjected either to scientific examination or to experiment. In this attitude there is more than a germ of truth. "Occam's razor" is as valuable an implement to-day as it ever was, and everyone will admit that we must exhaust all known causes before we proceed to postulate a new one.

We have gone beyond the day of the absurd statement that thought (which is of course unextended) is as much a secretion of the brain as bile (which, equally of course, is extended) is of the liver. No one nowadays would commit himself to such a statement, and men in general would be chary of urging that we should not believe anything which we cannot understand. I have myself heard a distinguished man of science of his day—he is dead this quarter of a century—make that statement in public, wholly ignoring the fact that any branch of science which we may pursue will supply us with a hundred problems we can neither understand nor explain, yet the factors of which we are bound to admit. But there is undoubtedly a dislike to accepting anything which cannot be proved by scientific means, and a tendency to describe as "mysticism"—a terrible and damning term to apply to anything, so its employers think!—any explanation which postulates something more in the universe than operations of a physical and chemical character.

My own opinion is that the state of things which we are considering finds its explanation in history, and I propose to devote a short space to [Pg 30] developing this view. Of course we might, and in some ways should, go back to the Reformation and to the destruction of religion which then took place. Let us, however, pass from that period to a time some hundred and fifty years ago and commence our investigations there, and in carrying them out I propose to make considerable use of the novels of different periods.

It is a truism that very little but the dry bones of history can be learnt from histories.

Nowadays people are sick of reading about more or less immoral monarchs, and more or less corrupt politicians, and it may be suspected that most of us have had our bellyful of wars now that the recent contest has come to an end. What one really wants to learn from history is how the ordinary folk, like ourselves, were getting on; what their ideas were; how the world wagged for them. Such information we are much more likely to get from memoirs and, since such works have been published, from novels. The novelist is not to be supposed to be committed to acceptance of all the remarks put into the mouths of his characters, but, if he is of the second, not to say the first flight (and, if he is not, he is not worth quoting), his characters and the general tone of his book will not be out of touch with the times to which they belong. Since the novel came into existence as something more than an occasional rarity, it is the novelists and not the players who are "the abstract and brief chronicles of the times," and it is to them that we shall apply for some of the information we desire.

[Pg 31]

To commence with the Georgian period, it is not too much to say that anything like real religion was scarcely ever at a lower ebb in England. This is not to say that there was an absolute dearth of religion. Law wrote his *Serious Call* during that period, and there are few books of its kind which have had a greater and more lasting effect. There were others of like but lesser character than Law, but, on the whole, no one will deny that the clergy of the Established Church (Catholics were, of course, in the catacombs) and the religion which they represented were almost beneath contempt. Look, for example, at *Esmond*, the typical novel of its period. Is there a single clergyman in it who is not an object of contempt, with the sole exception of the Jesuit, who, though a good deal of the stage variety, at least gains a measure of the reader's sympathy and respect? Thackeray was not himself a Georgian, it may be urged. That of course is true, but no one that knows Thackeray and knows also Georgian literature will deny that he was saturated with it and understood the period with which his book dealt better perhaps than those who lived in it themselves. But examine the novelists of the period; what about Fielding? Parson Adams is respectable and lovable, but the general average of parson and religion is certainly about as low as it can be. Fielding was not a religious man. Possibly, but what then of Richardson? We do not find religion at a very high level there; can anything well be more degraded than the figure cut by Mr. Williams in [Pg 32] *Pamela*, for example—the miserable curate upon whom the heroine calls for help in her distress? But apart from that, look at the whole atmosphere of the book. Why, the moral is that if you resist the immoral onslaughts of your master long enough he will give in and marry you, and you will be applauded for your successful strategy by all the countryside. Such is the book which all agreed to praise as an example of all that a book ought to be from the point of view of virtue.

It will be admitted by all conversant with the facts that religion could hardly have been at a lower ebb than it was when what is known as the Evangelical Movement came to trouble the placid, if stagnant and turbid, pool of the Established Church. Of course it did not transform the Church entirely. Read Miss Austen's novels: the most perfect pictures of life ever written. There are, I suppose, some half-dozen clergymen, pleasant and unpleasant, depicted in them, and we may be sure that they fairly well represent the typical average country parson of the period. Whatever they may otherwise be, they all agree in one point, namely in the complete absence of any such thing as a trace of spirituality. But in the early nineteenth-century Evangelicanism—specially that terrible variety Calvinism—was the dominant factor where religion really prevailed as a living influence; and it is to its influence, I firmly believe, that we may attribute the genuine detestation of religion which was so marked a feature of a part of the Victorian and most of the succeeding time. I am not, of course, for [Pg 33] getting the Oxford Movement, but, important as that was and is, in its earlier years it was almost entirely confined to clerical circles, exercising comparatively little influence on the laity and practically none at all on that great middle class which had been so much affected by the Wesleys, Whitefield, Scott, Newton, and the other pundits of Evangelicanism. Take the characteristic novel of the movement, if novel it should be called, Newman's *Loss and Gain*: I do not remember a single male character in it who is not in Holy Orders or on the way thereto. Hence, so far as religious influences are concerned, it is to the Evangelical Movement that we have to look. Now, though in my opinion it was the parent of many evils, there is no doubt that there was in it real fervour; intense devotion; a genuine desire to know and do God's will; a burning love for our Lord; coupled with all which were the most distorted and distorting ideas of what was and what was not sin ever conceived by any brain. Of this creed I can speak from personal knowledge, for I was brought up in it and know it from bitter experience.

The exponents of these views were never tired of instilling into their pupils the need for conversion, which was supposed to be a sudden operation. I have heard persons name the exact moment by the clock and the day on which theirs took place, and it was often effected by a single text. I have seen the Bible of an eminent leader in this line which contains a number of texts painted round with colours, each of which was [Pg 34] associated with the conversion of some particular individual. The process was supposed to be effected by the "acceptance of Christ," and though it was said to be free to all, it was clear to some at least of those who quite earnestly and really desired it, that, however ardent their desires, they could not secure their realisation. One was supposed to know in some mysterious manner that one was converted; the operation was permanent in its character; it could not be repeated; once thoroughly effected the converted person neither wished to sin nor really did sin. If anyone supposed to have been converted did relapse into evil ways, then he never had really been converted, but only seemed to have been. I have heard this circular form of argument urged most strongly by those who were (by constitution apparently) absolutely unable to see the illogical position which they were taking up. A further, and the most awful, part of the teaching was that however much one desired to be converted, and however earnestly one prayed for it, if one died without it damnation was certain. Lastly there was the encouraging thought that everything done prior to conversion was equally without merit; in fact, one might almost say, equally evil. These things were dinned into the heads of the young, in season and out of season; is it any wonder that so many of them grew up to hate religion? I remember myself the positive terror with which I went out even to minor entertainments, because I knew that in all probability close [Pg 35] interrogation would be made as to my spiritual condition.

Let me be reminiscent and recall one case. I was a boy at school and spending my Easter vacation away from home and with friends. It was my lot to have to dine one night with an old friend of my father's, a person of some distinction, who having, I believe, been a *viveur* in his youth, had in later years embraced the most ferocious type of Evangelicanism. When the ladies had retired I was left alone with this formidable person, whom I eyed much as a rabbit eyes a snake into whose cage he has been introduced. Nor were my fears groundless, for no sooner was the room empty than he peremptorily demanded of me whether I was saved. On hearing my trembling but perfectly truthful reply that I really did not know, he struck the table with his fist (I can see the whole thing quite plainly to-day, though it is five-and-forty years ago), exclaiming, "Then you are a fool, and if you were to die to-night you most certainly would be damned." I ask those who were brought up in a more kindly and more rational scheme of Christianity whether it is any wonder that those whose youth was spent in these gloomy shades should welcome the thought that there was no such being as a God?

Associated with this gloomy creed a new series of sins was invented, as if there were not enough already in the world. It was sinful to dance, even under the most domestic and proper circumstances. It was a sin to play cards, even [Pg 36] when there was no money on the game. It was a sin to go to the theatre, even to behold the most inspiring and instructive plays. It was even held by some, as we shall see, that the writing of stories or works of imagination was sinful. I once heard a professor of this creed express the doubt whether Shakespeare had not, on the whole, done much more harm than good, and state that he himself would not allow the works of Dickens to occupy a place in a hospital library, from which, as a matter of fact—for on this point the discussion had arisen—they had been excluded by the then chaplain of the institution, a man of like views. In fact, the idea of God which was presented to the youth of that period and brought up under such influences was—I do not say wilfully—that of a kind of super-policeman: a hard-hearted policeman, with an exaggerated code of misdoings, forever waiting round a corner to pounce on evil-doers, and, one was obliged to think, apparently almost pleased at the opportunity of catching them. It need not be said that no disrespect is intended in this. It is a simple and truthful statement of the kind of impression made upon one person by the teachings of that age and school. Is it any wonder that persons brought up in such a creed should experience a feeling of relief on learning that there was no God, no sin, no punishment? Add to this the terrors of the exaggerated Sabbatarianism of the period. What was the Sunday programme? Two lengthy sessions of Family Prayers; two attendances—each lasting at least an hour and a [Pg 37] quarter—on services in church; one, sometimes two, hours of Sunday School; no books but those of a religious character; no amusements of any kind even for the very young, unless the putting together of a dissected map of Palestine could be called an amusement; what a method of rendering Sunday attractive to the young!

Is it any wonder that those brought up on such a plan abandoned, with a sigh of relief, all religious exercises when at last they were able to do so? I notice that Mr. Belfort Bax, in his *Reminiscences of a Mid and Late Victorian*, alludes to this matter, saying that, "The most cruel of all the results of mid-Victorian religion was, perhaps, the rigid enforcement of the most drastic Sabbatarianism. The horror of the tedium of Sunday infected more or less the whole of the latter portion of the week." *Experto crede!* He says further, dealing with the 'fifties, that "the intellectual possibilities of the English people were then stunted and cramped by the influence of the dogmatic Calvinistic theology which was the basis of its traditional sentiment;"—it is exactly the point which I am trying to make.

We may now examine two instances of the kind of teaching with which I am dealing and its results. The first is that of the poet Cowper, and anyone who takes the trouble to read his life as written by Southey will find the whole piteous tale fully drawn out. Southey hated the Catholic Church, of which, by the way, he knew absolutely nothing, but he had sufficient sense to reject the teachings of Calvinism. [Pg 38] Cowper was at times insane and at other times of anything but a well-balanced mind, and he was just the kind of man who never ought to have been brought under the influences to which he was subjected. His principal adviser was the Rev. John Newton, a well-known Calvinistic clergyman of the Church of England. He must have been a man of compelling character, for he it was who brought the Rev. Thomas Scott, Rector of Aston Sandford, out of Socinianism, which, though a minister of the Church of England, he professed, into the Calvinistic view of things, as Scott himself tells us in his book *The Force of Truth*; and it must not be forgotten that it was to the writings of this same Scott that Newman tells us (in his *Apologia*) that he owed his very soul. Newton, like many of his fellows, had no sort of doubt as to his right to act as a director of souls, nor of his profound knowledge of how they should be dealt with. Yet it is to be remembered that, whilst the Catholic priest is obliged to undergo a long and careful training before he is permitted to take up this perilous task, Newton and those of his kind undertook it without any training whatever. Cowper, as everybody knows, was carefully and kindly tended by Mrs. Unwin, a woman a good deal older than himself, against whose character no word of reproach was ever uttered, the widow of an old friend of the poet. Newton wanted to drive Mrs. Unwin out of his house, but here at least Cowper rebelled and showed his very just annoyance, Newton actu [Pg 39] ally urged Cowper to abandon the task of translating Homer, a labour undertaken to distract his poor sick mind from thinking of itself, because such work, not being of a religious character, partook of the nature of sin. It is no wonder that such a rule of life had not infrequently the most distressing consequences. Newton himself admits that his preaching had the reputation of driving people into lunacy. In a letter asking that steps may be taken to remove one poor victim to an asylum he says: "I hope the poor girl is not without some concern for her soul; and, indeed, I believe a concern of this kind was the beginning of her disorder. I believe," he continues, "my name is up about the county for preaching people mad ... whatever may be the immediate cause, I suppose we have near a dozen, in different degrees, disordered in their heads, and most of them I believe truly gracious people."

Let us turn to the other example which I propose to select, that given by Mr. Gosse in his truly remarkable work *Father and Son*, one of the most faithful pictures of life ever written. The first instance shall be an extract from the diary of the mother, obviously a woman of great power and gifts if she had been given an opportunity of displaying them. "When I was a very little child," she writes, "I used to amuse myself and my brothers with inventing stories such as I had read. Having, as I suppose, naturally a restless mind and busy imagination, this soon became the chief pleasure of my life. Unfortunately my [Pg 40] brothers were always fond of encouraging this propensity, and I found in Taylor, my maid, a still greater tempter. I had not known there was any harm in it, until Miss Shore" (a Calvinistic governess), "finding it out, lectured me severely, and told me it was wicked. From that time forth I considered that to invent a story of any kind was a sin. But the desire to do so was too deeply rooted in my affections to be resisted in my own strength," (she was at this time nine years of age), "and unfortunately I knew neither my corruption nor my weakness, nor did I know where to gain strength. The longing to invent stories grew with a violence; everything I heard or read became food for my distemper. The simplicity of truth was not sufficient for me; I must needs embroider imagination upon it, and the folly, vanity and wickedness which disgraced my heart, are more than I am able to express. Even now (at the age of twenty-nine), though watched, prayed and striven against, this is still the sin which most easily besets me. It has hindered my prayers and prevented my improvement, and therefore has humbled me very much." It is narrated of the well-known Father Healy that a young lady having consulted him as to the sin of vanity, she feeling convinced, when she looked in her glass, that she was a very pretty girl, was answered by him, "My child, that is not a sin; it is a mistake!" It wanted some wise adviser to make the same remark to this poor tortured and deluded woman.

Illness under this code was always a punish [Pg 41] ment sent from heaven, as, indeed, it may be; but, "if anyone was ill it showed that 'the Lord's hand was extended in chastisement,' and much prayer was poured forth in order that it might be explained to the sufferer, or to his relations, in what he or they had sinned. People would, for instance, go on living over a cesspool, working themselves up into an agony to discover how they had incurred the displeasure of the Lord, but never moving away." One last instance, the most remarkable of all, and we may leave this book. It need hardly be said that a father of the kind depicted in this book would have a holy horror of the Catholic Church, and he had. He "welcomed any social disorder in any part of Italy, as likely to be annoying to the Papacy." He "celebrated the announcement in the newspapers of a considerable emigration from the Papal dominions, by rejoicing at this outcrowding of many, throughout the harlot's domain, from her sin and her plagues," and he even carried his hatred so far as to denounce the keeping of Christmas, which to him was nothing less than an act of idolatry.

On a certain Christmas Day, the servants, greatly daring, disobeyed the order of their master and actually had the audacity to make a small plum-pudding for themselves. Actuated by pity, no doubt, and by a feeling of kindness towards a small boy deprived of all the joys of the season, they pressed a slice of this pudding upon the son, who succumbed — very naturally — to the temptation. Shortly after, however, being afflicted [Pg 42] by a stomach-ache, remorse came upon him and he rushed to his father, exclaiming: "Oh! papa, papa, I have eaten of flesh offered to idols!" When the father learned what had happened, he sternly said, "Where is the accursed thing?" Having heard that it was on the kitchen table, "he took me by the hand, and ran with me into the midst of the startled servants, seized what remained of the pudding, and with the plate in one hand and me still tight in the other, ran till we reached the dust-heap, where he flung the idolatrous confectionery on to the middle of the ashes, and then raked it deep down into the mass. The suddenness, the velocity of this extraordinary act, made an impression on my memory which nothing will ever efface." Such is a plain unvarnished account of the kind of way in which numbers of people were brought up in the 'fifties and 'sixties of the last century. Can it be wondered that those who had such a childhood should grow up with an absolute horror of the Person in Whose name such things — absurdities when not positive crimes — were perpetrated? I firmly believe that these wholly false ideas of God and of sin have had more to do with the spread of materialism than many will perhaps be disposed to admit. Educated people, especially those trained in scientific methods, demand a certain common sense and sobriety in their beliefs. If they are brought up to believe that a grievous sin is committed when they invent an innocent story; when they go to a theatre or to a dance, or play a game of cards; if they have never [Pg 43] known the demands of real Christianity as put forward by the Catholic Church, is it likely that they will cleave to a faith which apparently engenders such absurdities as the Christmas pudding episode? It is, indeed, as Father Wasmann says, a thousand pities that the reasonableness, the logic, the dignity of the Catholic religion should remain for ever hidden from the eyes and minds of many who so often are as they are, because they were brought up as they were. In all these things we find the key to another problem. In another essay in this volume I have called attention to the glad intelligence, as it seems to a certain school of writers, that we are freed from the "bugbear of sin," as one of them puts it; able to enjoy ourselves without any thoughts of that kind.

Now I cannot but believe that such writers are thinking of the bugbear of artificial sins invented by the professors of a gloomy creed of religion. It is not to be supposed that any serious writer — and those to whom I allude are eminently such — would speak or write with pleasure and satisfaction of escaping from the bugbear of sins against morality or against one's neighbour; from the bugbear of dishonesty or theft; of taking away a person's character; of running away with his wife. I am convinced that it is the invented crimes of card-playing, theatre-going, and the like to which they are alluding: it could not surely be otherwise; and that makes it all the more unfortunate that before misusing a technical term like the word "sin," and thus perhaps mis [Pg 44] leading some young and ardent mind, such writers could not follow Father Wasmann's advice and study some simple manual of Catholic ethics, from which they would learn the real doctrine of Christianity and would discover how very different a thing it is and how very much more reasonable than the distorted caricature which we have been studying.

§ 2. THEOPHOBIA: ITS NEMESIS

Whether my view as to the cause, or one of the causes, is right or not, the fact remains that by the mid-Victorian period England had fallen to a very large extent a prey to materialism. Many people attribute the sudden onslaught of this to the publication of *The Origin of Species* and the controversies of the foolish which followed thereon. Samuel Butler, that brilliant writer who has not even yet come into his own, sums up in his novel *The Way of All Flesh* (and it may incidentally be remarked, in himself) most of the characteristics of the day. Many a parsonage home like that of the Rev. Theobald Pontifex existed in those days, and more than one Ernest Pontifex emerged from them. Now in this book Butler states that "the year 1858 was the last of a term during which the peace of the Church of England was singularly unbroken," and there no doubt he is right; "The Evangelical Movement ... had become almost a matter of ancient history. Tractarianism had subsided into a tenth-day's wonder; it was at work, but it was not noisy." [Pg 45] Then he says the calm was broken by the publication of three books: *Essays and Reviews, The Origin of Species, Criticisms on the Pentateuch* by Colenso. Few persons probably now remember the first and the last of these books; the fame of the second is likely to last long.

Whether again Butler is right in his idea as to the causes or not, as to the fact there can be no doubt. We have arrived at a period when the prevalent opinion amongst the intellectual classes was that religion—belief in anything which could not be fully understood—was impossible once one began to think seriously about it. Those who did not really look into such questions might go on considering themselves to believe in revelation, but the moment that a man seriously tackled the subject, his religion was bound to go, just as that of Ernest Pontifex did at the end of five minutes' conversation with an atheistic shoemaker. [21] Agnosticism and materialism were in the air, and remained the dominant features for quite a number of years. There were those who deplored the loss of their faith such as it had been. Huxley obviously did; and Romanes, who afterwards returned to the Church of England, con [Pg 46] fessedly did. Such persons, and there were many of them, honestly were unable to believe, and said so. A great deal of this was due to the attitude of popular science at that time. It was in a hot fit, and was going to explain everything, if not to-day, at least to-morrow. Now, as Sir Oliver Lodge told us before the war, in his book *Continuity*, we are in a cold fit and we seem only to know that nothing can be known. Sir Arthur Conan Doyle, best known as the creator of *Sherlock Holmes*, tells us in a recent book from which I shall have further to quote (*The New Revelation*, Hodder and Stoughton, 1918): "When I had finished my medical education in 1882, I found myself, like many young medical men, a convinced materialist as regards our personal destiny." With the facts contained in this statement I fully agree. The date in question is almost exactly that at which I also became a qualified medical man, and I, and I fancy most of my generation, believed ourselves to be agnostics if not atheists. It was the atmosphere of the time, and so strong as with difficulty to be resisted by those who resorted to the Universities. The point which I want to make is that during the latter part of the Victorian period we had come to a generation of intellectuals practically devoid of religion and followed in that respect by that always larger portion of any generation which, not having brains to think for itself, yet desiring to follow the intellectual *motif* of the day, adopts whatever is the fashionable attitude for the moment towards unseen things. Yesterday it was [Pg 47] blank negation; to-day it tends, as we shall see, to be spiritualism; to-morrow it might be earnest faith: let us hope so. And as to Calvinism, all this was *post hoc* of course; *propter hoc* also as I think.

What followed? That is what we now have to consider. The first thing which happened was the very natural discovery that science cannot explain everything; has in fact a strictly limited range of country to deal with. This discovery began to sap the foundations of materialism. Then there came the further discovery that all was not well, as so many supposed that it would be, under a scheme of life divorced from all connection with religion. Mr. Lucas, who has given the world many pleasant books, none of them with any obvious bias in favour of religion, in *Over Bemertons* (one of the most pleasant) makes one of his characters, *Mr. Dabney*, deplore the loss of the seriousness of the Victorian era: "We believe only in pleasure and success; our one ideal is getting wealth." Parenthetically, is not that just what might be expected? If there is really nothing but this world, what better can we seek than as much pleasure as we can get out of it? *Over Bemertons* was first published in 1908, and the remedy which *Mr. Dabney* then suggested, with a really curious prophetical insight, has just been vigorously applied. That remedy was "War, nothing more or less. A bloody war—not a punitive expedition or 'a sort of a war'" (he quoted these words with white fury) "'that might get us right again.' 'At great cost,' [Pg 48] I said. 'A surgical operation,' he replied, 'if the only means of saving life, cannot be called expensive.'"

Finally the discovery was made that mankind will not for long be content to do altogether without religion; a need for something more than bread alone being ingrained in his nature. Thus even the professedly materialistic societies try to afford something in the way of religious exercises. I have recently seen a notice of one of the so-called Ethical Societies in which the members (at their meetings, I take it) are "requested to silently meditate for five minutes on the good life." [22] It would seem to be quite as beneficial and more practical to meditate on split infinitives. A substitute for religion has to be found; what is it to be? In the years before the war Mr. Masefield published a very interesting book called *Multitude and Solitude*, which narrates the trials and troubles of two young Englishmen who make a perilous journey to Africa in search of the secret of the sleeping-sickness. In all their trials they never seem to have thought of prayer, in which it may be assumed they did not believe, but when they returned to England it occurred to one of them that there was something wanting in their life, and he propounded to his friend the view that "the world is just coming to see that science is not a substitute for religion," which is one of the things urged in this paper. He then proceeded to the [Pg 49] rather startling conclusion that science *is* "religion of a very deep and austere kind." One is reminded of a well-known passage in the Bible: "*Inveni et aram in qua scriptum erat* Ignoto Deo." To set up science as an "unknown God" seems a curious choice, even more curious than the choice of humanity, which—pitiable object as it is—was at least made in the image of God. Not to pile up instance upon instance, let us content ourselves with remembering that Mr. Wells, who in his earlier novels had certainly not displayed any marked affection for religion, in the last published before the war (*Marriage*) brings his hero face to face with the great realities, and makes him exclaim to his wife that he may "die a Christian yet," and urge upon her the need for prayer, if only out into the darkness. Of course, as all the reading world knows, since the war commenced, Mr. Wells has set up his own altar "Ignoto Deo," not with much more satisfactory results than those attained by Mr. Masefield. It is an historical fact that times of war have also been times of religious awakening, and it is natural that they should be so, for even the most careless must be brought to contemplate something more than the day's enjoyment. It is not then wonderful that the terrible war which has raged with Europe as the cockpit, and practically all the nations of the world as participants, should turn the minds of those who are in the righting line towards thoughts which in times of peace may never have found entrance there. From all sides one hears that this is so, yet here [Pg 50] again it is too often the case that an "unknown God" is sought, and from want of proper direction not always found. In a recently published memoir of one of the many splendid young fellows by whose death the world has been made poorer during this calamitous war, there is this moving passage: "I know that many hearts are turning towards *something*, but cannot find satisfaction in what the Christian sects offer. And many, failing to find what they need, fall back sadly into vague uncertainties and disbelief, as I often do myself." We badly need a St. Paul who will say to these and other anxious hearts, "*Quod ergo ignorantes colitis, hoc ego annuntio vobis.*"

However, it is much more with those who only "stand and wait" than with those who were actually in the trenches that we are concerned; what about the lamentable army of wives and mothers, widows and orphans, people bereft of those they loved or rising every morning in dread of the news which the day might bring forth; what about these and their attitude towards the things unseen? That many such have turned to some genuine form of religion is happily beyond dispute, but it is also unquestionably true that thousands have turned aside to the attractions of spiritualism. A recent article in the Literary Supplement of the *Times* commenced with the statement that "Among the strange, dismaying things cast up by the tide of war are those traces of primitive fatalism, primitive magic, and equivocal divination which are within general knowledge." The writer of the article in question [Pg 51] thinks that as we have taken a huge and lamentable step backwards in civilisation, we need not be surprised that we should also have receded in the direction of those primitive instincts to which he calls attention. This process had, however, begun long before the war.

The late Dr. Ryder, Provost of the Birmingham Oratory, was a very shrewd observer of public affairs and a very close and dear friend of the present writer. It must be more than twenty years ago since he remarked to me that he thought that materialism had shot its bolt and that the coming danger to religion was spiritualism, a subject on which, if I remember right, he had written more than one paper. I asked him what led him to that conclusion, and his reply was to ask me whether I had not noticed the great increase in number of the items in second-hand book catalogues—a form of literature to which we were both much addicted—under the heading "Occult." Since the war, however, there can be no doubt about the fact that spiritualism has made great strides. A thousand pieces of evidence prove it. Look, for example, at the enormous vogue of *Raymond*, a book of which I say nothing, out of personal regard for its author and genuine respect for his honesty and fearlessness. But I return to Sir Arthur Doyle's book, and we find him assuring us that he is personally "in touch with thirteen mothers who are in correspondence with their dead sons," and adds that in only one of these cases was the individual concerned with psychic matters before [Pg 52] the war. Further, he explains that it was the war which induced him to take an active interest in a subject which had been before no more than one of passing curiosity. "In the presence of an agonised world," he writes, "hearing every day of the deaths of the flower of our race in the first promise of their unfulfilled youth, seeing around one the wives and mothers who had no clear conception whither their loved one had gone to, I seemed suddenly to see that this subject with which I had so long dallied was not merely a study of a force outside the rules of science, but that it really was something tremendous, a breaking down of the walls between the two worlds, a direct undeniable message from beyond, a call of hope and of guidance to the human race at the time of its deepest affliction." Perhaps it is not wonderful that spiritualism should have won the success which it has, for it offers a good deal to those who can believe in it. It offers definite intercourse with the departed; positive knowledge as to the existence of a future state, and even as to its nature—the last-named intelligence not always very attractive. Further, it requires no particular creed and, it would appear, no special code of morals; for one of its teachings, I gather, is that it does not greatly matter what a man thinks or even does, so far as his future welfare is concerned.

Sir A. Doyle's book is the least convincing exposition of spiritualism I have yet read—and I have studied many of them—but it may be taken to include the latest views on the subject. [Pg 53] Amongst the revelations which he gives, there is one purporting to come from a spirit who "had been a Catholic and was still a Catholic, but had not fared better than the Protestants; there were Buddhists and Mahommedans in her sphere, but all fared alike." Another spirit informed Sir A. Doyle that he had been a freethinker, but "had not suffered in the next life for that reason." This is not the occasion, and in no way am I the man, to tackle the subject of spiritualism, but this at least I think may be said, that the person who argues that the whole thing is a fraud and deception does not know what he is talking about. Look at the history of the world—*Quod semper, quod ubique*, almost *quod ab omnibus*. The records of early missionaries—Jesuits especially—teem with accounts of the same kind of phenomena as we read of in connection with séances to-day, occurring in all sorts of places and amongst widely separated races of mankind. We have it in the *Odyssey*; we have it in Cicero and in Pliny; we have it in the Bible. All this is not a mere matter of imposition.

In a very curious book recently published (*Some Revelations as to "Raymond,"* by a Plain Citizen; London, Kegan Paul), to which some attention may now be devoted, the writer, himself a firm believer in spiritualism and one obviously in a position to write about it, points out that the old term "magic" has been relegated to the performances of conjurers, and the terminology so altered as to make spiritualism appear to be a new gospel, whereas the contrary is the case. [Pg 54] "The impression prevailed that civilised people were in presence of a new order of phenomena, and were acquiring a new outlook into the regions of the Unknown; whereas the truth was that they were merely repeating, under new social conditions and in a new environment, the same experiences that had happened to their ancestors during some thousands of years." Here I may interject the remark that as far as my reading and knowledge go, no spirit has ever had a good word to say for the Catholic religion. What that Church thinks about spiritualism has been made quite clear, and that is enough for Catholics. Before leaving the Plain Citizen, we must not omit to notice one strange hypothesis of his, all the stranger as coming from a professed spiritualist. He maintains—perhaps it would be fairer to say that he lays down as a working hypothesis—the following thesis: Spiritualism involves the existence of mediums, and mediums for the most part have to make their living by their operations. They will not be averse to making their incomes as large as possible. For the purpose of acquiring information as to the affairs of possible clients, they have, so he asserts, an almost Freemasonic Association by which all sorts of pieces of intelligence concerning persons of importance are collected and disseminated amongst the brotherhood. It did not require much imagination to suppose that the war would add to the number of their clients, whether their claims had real foundation or not; what they wanted above all things was some one of undoubted position who would [Pg 55] "boom the movement," in the slang of the day. They laid all their plans to get their man in the author of *Raymond*, and they got him. Such is his thesis for what it is worth.

However, it is time to conclude. What I wanted to show was that Theophobia was the Nemesis of a dreadful type of Protestantism, and that spiritualism was the Nemesis of the materialism associated with that Theophobia. There is no need to point out to Catholic readers where the remedy lies, and where the real Communion of the saints is to be found. They are not likely to be drawn aside by the "Lo here!" of the "false Christs" whom we were promised and whom we are getting. It is for those who have themselves experienced the consolations of the Catholic religion to do their best, each in his own way, to make known to others outside our body what things may be found within.

FOOTNOTES:

[Pg 56]

[21] An excellent example may be found in Butler's own career. Destined for the ministry of the Church of England (with his own full consent), he was set to teach a class in a Sunday school. Finding that some of his pupils were unbaptized, yet no worse-behaved than the others, and obviously quite ignorant of what baptism meant, he abandoned all belief. His biographer, equally ignorant, in narrating, with approval, this change of opinion, says, "Paley had produced evidence of Christianity, but none so unmistakable as this to the contrary."

[22] Dr. Johnson once remarked that "to find a substitution for violated morality was the leading feature in all perversions of religion."

III. WITHIN AND WITHOUT THE SYSTEM

Exclusive and long-continued devotion to any special line of study is liable to lead to forgetfulness of other, even kindred, lines—almost, in extreme cases, to a kind of atrophy of other parts of the mind. There is the example of Darwin and his self-confessed loss of the æsthetic tastes he once possessed. Nor are scientific studies the only ones to produce such an effect. The amusing satire in *The New Republic* has, perhaps, lost some of its tang now that the prototype of its Professor of History is almost forgotten, but it has not lost its point. Lady Ambrose tells the tale: "He said to me in a very solemn voice, 'What a terrible defeat that was which we had at Bouvines!' I answered timidly—not thinking we were at war with anyone—that I had seen nothing about it in the papers. 'H'm!' he said, giving a sort of grunt that made me feel dreadfully ignorant, 'why, I had an excursus on it myself in the *Archæological Gazette* only last week.' And, do you know, it turned out that the Battle of Bouvines was fought in the Thirteenth Century, and had, as far as I could make out, something to do with Magna Charta."

[Pg 57]

It is, however, among writers on biological subjects that we find the most salient instances of this contraction. With extraordinary self-abnegation they seem, in the contemplation of the problem with which they are concerned, to forget that they themselves are living things, and, more than that, the living things of whom they ought to know and could know most, however little that most may be. When the biologist begins to philosophise as, after the manner of his kind, he often does, he should leave his microscope and look around him; whereas he often forgets even to change the high for the low power. Thus he limits his field of vision and forgets, when attempting his explanation, that it is only *within a system* that he is working. Professor Ward, in *Naturalism and Agnosticism*, says:

"From the strict premisses of Positivism we can never prove the existence of other minds or find a place for such conceptions as cause and substance; for into these premisses the existence of our own mind and its self-activity have not entered. And accordingly we have seen Naturalism led on in perfect consistency to resolve man into an automaton that goes of itself as part of a still vaster automaton, Nature as mechanically conceived, which goes of itself. True, this mechanism goes of itself because it *is* going, and being altogether inert, cannot stop or change. How it ever started is indeed a question which science cannot answer, but which, on the other hand, it has no occasion to ask: time, its one [Pg 58] independent variable, extends indefinitely without hint of either beginning or end. Such a system of knowledge, *once we are inside it*, so to say, is entirely self-contained and complete."

"*Once we are inside it!*" what so many writers forget or ignore is that they *are* inside it, and that their explanations do not explain the system or how it came to be there or to be in operation. Everybody is familiar with Paley's example of the watch found on the heath. Let us carry it a little further. Suppose some student, after devoting years of patient examination to the watch, were to come forward and say: "I have discovered the secret of this watch. There is a spring in it which possesses resiliency, and it is that which drives the wheels. I think I have heard people say that there must have been a watchmaker to design and construct this piece of machinery, but, in face of my discoveries, any such explanation is wholly unnecessary and may be altogether abandoned."

Perhaps this analogy may be regarded as exaggerated; but, before thus condemning it, let the following passage be studied. It is from a very important book recently published, which claims (and has had its claim supported by many periodicals) to have done away with any need for an explanation of life beyond that which can be given by chemistry and physics, Jacques Loeb's *Organism as a Whole, from a Physico-Chemical Viewpoint*.

It would be hard to find a worse example of [Pg 59] confused thinking than that of the following passage:

"The idea that the organism as a whole cannot be explained from a physico-chemical viewpoint rests most strongly on the existence of animal instincts and will. Many of the instinctive actions are 'purposeful,' *i.e.* assisting to preserve the individual and the race. This again suggests 'design' and a designing 'force,' which we do not find in the realm of physics. We must remember, however, that there was a time when the same 'purposefulness' was believed to exist in the cosmos where everything seemed to turn literally and metaphorically around the earth, the abode of man. In the latter case, the anthropo- or geo-centric view came to an end when it was shown that the motions of the planets were regulated by Newton's law, *and that there was no room left for the activities of a guiding power.* Likewise, in the realm of instincts, when it can be shown that these instincts may be reduced to elementary physico-chemical laws, the assumption of design becomes superfluous." (*Italics mine.*)

In the first place the "purposefulness" of the movements of the planets is not affected in the very least by the question of heliocentricism. What the author is probably thinking of is an exaggerated and obsolete teleology, but that is not what seems to be the purport of the passage. Let that pass. The main confusion [Pg 60] lies in the application of the term "Law." The Ten Commandments, and our familiar friend D.O.R.A., are laws we must obey or take the consequences of our disobedience. The "laws" which the writer is dealing with are not anything of this kind. Newton's Law is not a thing made by Newton, but an orderly system of events which was in existence long before Newton's time, but was first demonstrated by him. It tells us how a certain part of the system works—when we are *"inside it."* It does not in the least explain the system any more than the discovery of the resiliency of the spring of the watch explains the watch itself. So far from dispensing with "the activities of a guiding power," Newton's law is positively clamant for a final explanation, since it does not tell us, nor does it pretend to tell us, how the "law" came into existence, still less how the planets came to be there, or how they happen to be in a state of motion at all. Writers of this kind never seem to have grasped the significance of such simple matters as the different kinds of causes, or to be aware that a formal cause is not an efficient cause, and that neither of them is a final cause. Coming to the latter part of the paragraph, it is in no way proved that instincts can be reduced to physico-chemical laws, and, suppose it were proved, the assumption of design would be exactly where it is at this moment. It is the old story of St. Thomas Aquinas and Avicenna and their discussion on abiogenesis, and surely biologists might be expected to have heard of that. The same confusion of thought is to be met with [Pg 61] elsewhere in this book, and in other similar books, and a few instances may now be examined.

Samuel Butler, in *Life and Habit*, warns his readers against the dicta of scientific men, and more particularly against his own dicta, though he made no claim to be a scientist. If his reader *must* believe in something, "let him believe in the music of Handel, the painting of Giovanni Bellini, and in the thirteenth chapter of St. Paul's first Epistle to the Corinthians." And he exclaims: "Let us have no more 'Lo, here!' with the professor; he very rarely knows what he says he knows; no sooner has he misled the world for a sufficient time with a great flourish of trumpets than he is toppled over by one more plausible than himself." That is a somewhat unkind way of putting it; but undoubtedly theory after theory is put forward, and often claimed to be final, only to disappear when another explanation takes its place. Thus at the moment we are in the full flood of the chemical theory which is employed to explain inheritance. That heredity exists we all know, but so far we know nothing about its mechanism. Darwin, with "Pangenesis," and others, using other titles, argued in favour of a "particulate" explanation, but the number of particles which would be necessary to account for the phenomena involved, this and other difficulties, have practically put this explanation out of court. Then we had the Mnemic theory of Hering, Butler, and others, by which the unconscious memory of the embryo—even the germ—is the explanation. Quite lately the [Pg 62] mnemic theory has been claimed by Rignano in his *Scientific Synthesis* as a complete explanation, in forgetfulness of the fact that even the all-powerful protozoon can only remember what has passed and could certainly not *remember* that it was some day going to breed a man. At the moment, things are explained on a chemical basis, though that basis is far from firm; is of a shifting nature, and a little hazy in details. Some time ago, colloids were the cry. A President of the British Association almost led one to imagine that "the homunculus in the retort" might be expected in a few weeks. But the chemists would have none of this, and denied that the colloids, about which they ought to know more than do the biologists, had that promise in them which had been claimed. We had Leduc and his "fairy flowers," as now we have Loeb and others with their metabolites and hormones. As to these last, there seems to be no kind of doubt that the internal secretions of many organs and structures have effects which were, even a few years ago, quite unsuspected. Those of the thyroid and adrenals are excellent examples.

It seems to be the fate, however, of all supporters of new theories to run into extravagances. Darwin had to remind his enthusiastic disciples that Natural Selection could not create variations, and we may feel some confidence that Hering, were he alive, would urge his followers to bear in mind that memory cannot create a state of affairs which never existed. So far we may certainly say that these internal secretions do produce [Pg 63] certain physical effects, some of them effects not to be suspected by the uninformed reader. There seems to be very good evidence that the growth of antlers in deer depends upon an internal secretion from the sex-gland and from the interstitial tissue of that gland; for it is apparently upon the secretions of this portion of the gland that the secondary sexual characters depend, and not merely these, but also the normal sexual instincts. And this takes us a stage further. The extreme claim is that all instincts, in fact all thoughts and operations, are in the last analysis chemical or chemico-physical. Let us examine this claim for a moment. The adrenals are two inconspicuous ductless bodies situated immediately above the kidneys. Not many years ago, when the present writer was a medical student, all that was known about these organs was that when stricken with a certain disease, known as Addison's disease from the name of its first describer, the unfortunate possessor of the diseased glands became of a more or less rich chocolate colour. To-day we know that the internal secretion of these organs is a very powerful styptic, and there is good reason to believe that a copious discharge accompanies an unusual exhibition of rage. When we are told things of this kind we must first of all remember that the adrenalin does not cause the rage, though it may produce its concomitant phenomena. If a man flies into a violent passion because someone has trodden upon his corns, and there is a copious flow of adrenalin from the glands, it is not that [Pg 64] flow which has caused his rage. It may be the flow from the interstitial tissue of the sex-glands which engenders sexual feelings, but then those are almost wholly physical, and only in a very minor sense — if even if any true sense — psychical. Persons who take the extreme view have never yet suggested that there is a characteristic hormone connected with those psychical attributes alluded to in the chapter of the Corinthians recommended to our notice by Butler. In fact they seem to ignore all but the lower or vegetable characters when dealing with psychology from the chemico-physical point of view.

Finally, we come again to the fatal and fundamental defect of this as of other "explanations"; it is an explanation "*within the system*," and therefore unphilosophical in so far as it fails to explain the facts through their ultimate or deepest reasons.

A large part of Loeb's book is devoted to a description of the author's remarkable experiments in artificial parthenogenesis, and an attempt to show that they offer a complete explanation. Sir William Tilden, one of the greatest living authorities on organic chemistry, tells us that "too much has been made of the curious observations of J. Loeb and others"; and he definitely states that when we consider "the propagation of the animal races by the sexual process ... there can be no fear of contradiction in the statement that in the whole range of physical and chemical phenomena there is no ground for even a suggestion of an explanation." Behind [Pg 65] this pronouncement of an expert, one might well shelter oneself; but the question under consideration merits a little further treatment. The reproduction of kind, though usually a bi-sexual process, may, however, normally in rare cases be uni-sexual, and this process is known as Parthenogenesis. Even in human beings certain tumours of the sex-glands, known as teratomata, very rare in women and even rarer, if ever existent, in men, have been claimed as examples of attempts at parthenogenesis, and so far no better explanation is available.

Now Loeb and others have succeeded in certain forms—even in a vertebrate like the frog—in inducing development in unimpregnated ova. The evidence for all these things is still slender; but we will content ourselves with noting that point and passing on to the consideration of the phenomena and the claims put forward in connection with them. We find the task of unravelling the writer's meaning rendered more difficult by a certain confusion in his use of terms, since fertilisation, *i.e.* syngamy—the union of the different sex products—seems to be confused with segmentation, *i.e.* germination; and this confusion is accentuated by the claim that "the main effect of the spermatozoon in inducing the development of the egg consists in an alteration in the surface of the latter which is apparently of the nature of a cytolysis of the cortical layer. Anything that causes this alteration without endangering the rest of the egg may induce its development." When the spermatozoon enters [Pg 66] the ovum it causes some alteration in the surface membrane of the latter which, amongst other things, prevents the entrance of further spermatozoa. Loeb thinks that in causing this alteration it sets up the segmentation of the ovum. That there is a close connection between the two events seems undoubted; that they are in relation of cause and effect seems likely. It is quite evident that an artificial stimulus can in certain cases set up segmentation, but never can it cause the fertilisation of the ovum. It may very likely produce the same change in the membrane that is caused by the entrance of the spermatozoon under normal circumstances—membrane formation may be necessarily coincident with the liberation in the egg of some zymose which arises from a pre-existent zymogen. But we are still some way off any assurance that the *main* object of the spermatozoon in inducing the development of the egg is this surface alteration. It may be the initial effect; very probably it is; but since the main function of the spermatozoon must be the introduction of germplasm from the male parent, it is too much for anyone to ask us to believe that its *main* function is concerned with surface alteration.

Loeb argues that the change in the surface membrane is of a chemical character, and that no doubt may be correct; but even if we allow him every scientific fact, or surmise, he is still, as in the other cases with which we have dealt, miles away from any real explanation. He is still inside his chemico-physical explanation to begin [Pg 67] with; and, even within that, he still leaves us anxious for the explanation of a number of points—for example, as to the nature of the chemical process which accompanies, or is the cause of, segmentation. We in no way press these questions; for similar demands could be made in so many cases; we only indicate that they are there. What we do press is this—that when an authority comes forward to assure us that all the processes of life, including man's highest as well as his lowest attributes, can be explained on chemico-physical lines, we are entitled to ask for a more cogent proof of it than the demonstration, however complete, of the germination of an egg, caused by artificial stimulus and not by the ordinary method of syngamy, even though that germination may lead to the production of a perfect adult form. We are entitled to ask him to make clear to us not only what is happening *within his system*, but—which is far more important—what that system is, and how it came into existence. We are entitled to ask why the artificial stimulus, or the entry of the spermatozoon, produces the effects which it is claimed to produce instead of any one of some score of other effects which it might conceivably have produced. Above all we are entitled to ask why there are any effects, or even why there is any ovum or any spermatozoon or curious physiological investigator, to give the artificial stimulus. Until some light is thrown upon these things we are still within the system, or merely hovering round its confines, and are far away from any final or philo [Pg 68] sophical explanation such as would satisfy the mind of the man who wants to get a real and not a partial knowledge of the things around him.

We may now turn to the question of Vitalism. It was long the regnant theory; then temporarily the Cinderella of biology; it is now returning to its early position, though still denied by those of the older school of thought who cannot imagine the kitchen wench of yesterday the ruler of to-day. One of the objections to Vitalism is that this explanation of living things is thought by ignorant writers to be so inextricably mixed up with theological considerations as to furnish a case of *stantis aut cadentis ecclesiae*. That is, of course, absurd; but it creates an undoubted bias against the theory. Hence it is the fashion amongst its opponents to write of it as "mystical" or, as Loeb does, as "supernatural," probably the most illogical term that could possibly be used. What is Vitalism? It is the theory that there is some other element—call it entelechy with Driesch, or call it what you like—in living things than those elements known to chemistry and physics. If it is *not* there, *cadit quaestio*; if it *is* there it is not "supernatural." It might with reason be called "super-mechanical," or "super-chemical," or "super-physical"; but if it is in Nature, as it is held to be, it is not "supernatural" in any true sense of that word—no dictionary confines the term "Nature" to the operations of chemistry and physics.

A good deal of the misconception existing on this point comes from pure ignorance of philo [Pg 69] sophy, a subject with which writers of this school seldom have even a nodding acquaintance. "The idea of a quasi-superhuman intelligence presiding over the forces of the living is met with in the field of regeneration." Echoes of the Cartesian idea of the soul seem to ring in this statement; but it could not have been written by anyone who had mastered the Aristotelian or the Scholastic explanation of matter and form. But let us take this question of Regeneration; the power which all living things have, in some measure, though in very different measure, of reconstructing themselves when injured. It has been dealt with in a masterly manner by Driesch; and we may at once say that we do not think that Loeb has in any way controverted his argument, nor even entered the first line of defence of that which is built up around what he calls by the somewhat forbidding name of "Harmonious-Equipotential System."

Let us take one particular example, a very remarkable one, which has been cited by both writers—Wolff's experiment on the lens of the eye. The lens is just behind the pupil or central aperture in the iris or coloured ring at the front of the eye, and behind the cornea which is to the eye what a watch-glass is to a watch. If the lens of the eye be removed from a newt, as it is from human beings in the operation for cataract, the animal will grow another one. How does it do it? In certain cases a tiny fragment of the lens has been left behind after the operation, and the new one grows from that. This is sufficiently [Pg 70] wonderful, but by no means so wonderful as what happens in other cases in which the entire lens has been removed and the new lens grows from the outer pigmented layer of the margin of the iris. To the unbiological reader one source of origin will not seem more wonderful than the other, but there is really a vast distinction between them. At an early stage in the development of the embryo, the cells composing it become divisible into three layers. It is even possible, as Loeb maintains, that this differentiation is present in the unsegmented ovum, in which case the facts to be detailed become still more remarkable and significant. These layers are known as epi-, meso-, and hypo-blast; and from each one of them arise certain portions of the body, and certain portions only. It would be as remarkable to a biologist to find these layers not breeding true as it would to a fowl-fancier to discover that the eggs of his Buff Orpingtons were producing young turkeys or ducks. Now the lens is an epiblastic structure, and the iris is mesoblastic. Hence the wonder with which we are filled when we find the iris growing a lens. Loeb attempts to explain this in the first instance by telling us that the cells of the iris cannot grow and develop as long as they are pigmented; that the operation wounds the iris, allows pigment to escape, and thus permits of proliferation. We may accept this, and yet ask why it takes on a form of growth familiar to us only in connection with epiblast? The reply is: "Young cells when put into the optic cup always become transparent, no matter [Pg 71] what their origin; it looks as if this were due to a chemical influence, exercised by the optic cup or by the liquid it contains.

"Lewis has shown that when the optic cup is transplanted into any other place under the epithelium of a larva of a frog the epithelium will always grow into the cup where the latter comes in contact with the epithelium; and that the ingrowing part will always become transparent." A most remarkable and interesting experiment; it has this very important limitation—that it is always *epithelium* with which it has to do, whereas in Wolff's experiment the regeneration takes place from mesoblastic tissue. The cause of the transparency may be a chemical reaction—it depends a good deal upon our definition of that phrase. Is protoplasm a chemical compound? Some have considered it so, and spoken of its marvellously complicated molecule. Of course it is made up of carbon, hydrogen, and other substances within the domain of chemistry. But is it, therefore, merely a chemical compound? The reply involves the whole riddle of Vitalism. The author would say that it, as well as all the living things to which it belongs, is purely and solely a chemical compound; and he must take the consequences of his belief. One of these consequences, from which doubtless he would not shrink, would be that a super-chemist (so to speak) could write him and his experiments and his book down in a series of chemical formulæ—a consequence which takes a good deal of believing. But it also involves him in a belief in the rigidity [Pg 72] of chemical reactions; and we are entitled to ask for an explanation of the identical behaviour of the chemical reaction in connection with epiblastic and mesoblastic cells—both pure chemical compounds *ex hypothesi* and, as far as we can tell from their normal behaviour, widely differing from one another. The optic cup, or its contained fluid, is one chemical compound; epithelium is another; mesoblast is a third. We want an explanation of the identical behaviour of the first with *either* of the two latter; and this should be borne in mind—that the reaction is not a mere matter of "clearing" of a tissue as the histologist would clear his section by oil-of-cloves or other reagent, but of the construction of a different type of cell—epithelial, not connective tissue.

It certainly follows that there must be some superior, at least widely different, agency at work than one of a purely chemical character—something which transcends chemical operations. This is precisely what the Vitalist claims. No one will fail to award praise to any attempts to explain the phenomena of Nature, whether within or without any system. Loeb's book sets out to do a great deal more—to explain what it does not explain—the Organism as a Whole, and thus to give a philosophical explanation of man. It even claims to afford hints for a rule for his life, at least so we gather from the Preface, where, alluding to "that group of freethinkers, including d'Alembert, Diderot, Holbach and Voltaire," the author tells us that they "first dared to follow the consequences of a mechanistic science—incomplete as [Pg 73] it then was—to the rules of human conduct, and thereby laid the foundation of that spirit of tolerance, justice, and gentleness which was the hope of our civilisation until it was buried under the wave of homicidal emotion which has swept through the world." On which it is surely reasonable to ask how a chemical reaction can learn so to alter itself as to exhibit "tolerance, justice, and gentleness," attributes which it had not previously possessed? Such claims of this and other writers, who would find in the laws of Nature as formulated to-day (forgetful that their formulæ may to-morrow be cast into the furnace) a rule of life as well as a full explanation of the cosmos, resemble in their lack of base an inverted pyramid.

[Pg 74]

IV. SCIENCE IN "BONDAGE"

Amongst the numerous taunts which are cast at the Catholic Church there is none more frequently employed, nor, it may be added, more generally believed, nor more injurious to her reputation amongst outsiders—even with her own less-instructed children themselves at times—than the allegation which declares that where the Church has full sway, science cannot flourish, can scarcely in fact exist, and that the Church will only permit men of science to study and to teach as and while she permits.

To give but one example of this attitude towards the Church, readers may be reminded that Huxley [23] called the Catholic Church "the vigorous enemy of the highest life of mankind," and rejoiced that evolution, "in addition to its truth, has the great merit of being in a position of irreconcilable antagonism to it." An utterly incorrect, even ignorant statement, by the way—but let that pass. The same writer, in a number of places, in season and out of season, as we may fairly say, [24] proclaims his wholly erroneous view that there is "a necessary antagonism between [Pg 75] science and Roman Catholic doctrine." We need not labour this point. It is sufficiently obvious, nor does it need any catena of authorities to establish the fact, that outside the Church, and even, as we have hinted above, amongst the less-instructed of her own children, there is a prevalent idea that the allegation with which this paper proposes to deal is a true bill.

Those who give credit to the allegation must of course ignore certain very patent facts which are, it will be allowed, a little difficult to get over. They must commence by ignoring the historical fact that the greater number—almost all indeed—of the older Universities, places specially intended to foster and increase knowledge and research, owe their origin to Papal bulls. They must ignore the fact that vast numbers of scientific researches, often of fundamental importance, especially perhaps in the subjects of anatomy and physiology, emanated from learned men attached to seats of learning in Rome, and this during the Middle Ages, and that the learned men who were their authors quite frequently held official positions in the Papal Court. They must finally ignore the fact that a large number of the most distinguished scientific workers and discoverers in the past were also devout children of the Catholic Church. Stensen, "the Father of Geology" and a great anatomical discoverer as well, was a bishop; Mendel, whose name is so often heard nowadays in biological controversies, was an abbot. And what about Galvani, Volta, Pasteur, Schwann (the originator of the Cell Theory), van Beneden, [Pg 76] Johannes Müller, admitted by Huxley to be "the greatest anatomist and physiologist among my contemporaries"? [25] What about Kircher, Spallanzani, Secchi, de Lapparent, to take the names of persons of different historical periods, and connected with different subjects, yet all united in the bond of the Faith? To point to these men—and a host of other names might be cited—is to overthrow at once and finally the edifice of falsehood reared by enemies of the Church, who, before erecting it, might reasonably have been asked to look to the security of their foundations.

Still there is the edifice, and as every edifice must rest on some kind of foundation or another, even if that foundation be nothing but sand, it may be useful and interesting to inquire, as I now propose to do, what foundation there is—if in fact there is any—for this particular allegation.

We might commence by interrogating the persons who make it. The probability is that the reply which would at once be drawn from most of them would amount to this: "Everybody knows it to be true." If the interrogated person is amongst those less imperfectly informed we shall probably be referred to Huxley or to some other writer. Or we may even find ourselves confronted with that greater knowledge—or less inspissated ignorance—which babbles about Galileo, the Inquisition, the *Index*, and the *imprimatur*.

Galileo and his case we shall consider later on, [Pg 77] for he and it are really germane to the question with which we are dealing. The Inquisition has really nothing to do with the matter. The *Index* we also reserve for a later part of this essay. With the *imprimatur* we may now deal, since there is no doubt that there is a genuine misunderstanding on this subject on the part of some people who are misled perhaps through ignorance of Latin and quite certainly through ignorance of what the whole matter amounts to. Let us begin by reminding ourselves that, though the unchanging Church is now, so far as I am aware, the only body which issues an *imprimatur*, there were other instances of the exercise of such a privilege even in recent or comparatively recent days. There were Royal licences to print with which we need not concern ourselves. But, what is important, there was a time when the scientific authority of the day assumed the right of issuing an *imprimatur*. I take the first book which occurs to me, Tyson's *Anatomie of a Pygmie*, and for the sake of those who are not acquainted with it, I may add that this book is not only the foundation-stone of Comparative Anatomy, but also, through its appendix *A Philological Essay Concerning the Pygmies, the Cynocephali, the Satyrs, and Sphinges of the Ancients*, the foundation-stone of all folk-lore study. On the page fronting the title of this work the following appears:

17 Die Maij, 1699.

Imprimatur Liber cui Titulus, Orang-Outang sive Homo Sylvestris, etc. Authore Edvardo Tyson, M.D., R.S.S.

John Hoskins, V.P.R.S.

[Pg 78]

What does this mean? In the first place it shows, what all instructed persons know, that the Royal Society did then exercise the privilege of giving an *imprimatur* at any rate to books written by its own Fellows. It cannot be supposed that such *imprimatur* guaranteed the accuracy of all the statements made by Tyson, for we may feel sure that John Hoskins was quite unable to give any such assurance. We must assume that it meant that there was nothing in the book which would reflect discredit upon the Society of which Tyson was a Fellow and from which the *imprimatur* was obtained.

However this may be, the sway over its Fellows' publications was exercised, and indeed very excellent arguments might be adduced for the reassumption of such a sway even to-day. [26]

Though the *imprimatur* in question has fallen into desuetude, it is, as we all know, the commonest of things for the introductions to works of science to occupy some often considerable part of their space with acknowledgments of assistance given by learned friends who have read the manuscript or the proofs and made suggestions with the object of improving the book or adding to its accuracy. Any person who has written a book can feel nothing but gratitude towards those who have helped him to avoid the errors and slips to which even the most careful are subject.

[Pg 79]

So that such acknowledgments of assistance have come to be almost what the lawyers call "common form." What they really amount to is a proclamation on the part of the author that he has done his best to ensure that his book is free from mistakes. Now the *imprimatur* really amounts to the same thing, for it is, of course, confined to books or parts of books where theology or philosophy trenching upon theology is concerned. Thus a book may deal largely, perhaps mainly, with scientific points, yet necessarily include allusions to theological dogmas. The *imprimatur* to such a book would relate solely and entirely to the theological parts, just as the advice of an architectural authority on a point connected with that subject in a work in which it was mentioned only in an incidental manner, would refer to that point, and to nothing else. Perhaps it should be added, that no author is obliged to obtain an *imprimatur* any more than he is compelled to seek advice on any other point in connection with his book. "*Nihil Obstat*," says the skilled referee: "I see no reason to suppose that there is anything in all this which contravenes theological principles." To which the authority appealed to adds "*imprimatur*:" "Then by all means let it be printed." The procedure is no doubt somewhat more stately and formal than the modern system of acknowledgments, yet in actual practice there is but little to differentiate the two methods of ensuring, so far as is possible, that the work is free from mistakes. That neither the assistance of friends nor the *imprimatur* of [Pg 80] authorities is infallible is proved by the facts that mistakes do creep into works of science, however carefully examined, and that more than one book with an *imprimatur* has, none the less, found its way on to the *Index*. Before leaving this branch of the subject one cannot refrain from calling attention to another point. How often in advertisements of books do we not see quotations from reviews in authoritative journals—a medical work from the *Lancet*, a physical or chemical from *Nature*? Frequently too we see "Mr. So-and-So, the well-known authority on the subject, says of this book, etc., etc." What are all these authoritative commendations but an *imprimatur* up to date?

Passing from the *imprimatur* to a closer consideration of our subject, it is above all things necessary to take the advice of Samuel Johnson and clear our minds of cant. Every person in this world—save perhaps a Robinson Crusoe on an otherwise uninhabited island, and he only because of his solitary condition—is in bondage more or less to others; that is to say, has his freedom more or less interfered with. That this interference is in the interests of the community and so, in the last analysis, in the interests of the person interfered with himself, in no way weakens the argument; it is rather a potent adjuvant to it. However much I may dislike him and however anxious I may be to injure him, I may not go out and set fire to my neighbour's house nor to his rick-yard, unless I am prepared to risk the serious legal penalties which will be my lot if I am detected [Pg 81] in the act. I may not, if I am a small and active boy, make a slide in the public street in frosty weather, unless I am prepared—as the small boy usually is—to run the gauntlet of the police. In a thousand ways my freedom, or what I call my freedom, is interfered with: it is the price which I pay for being one item of a social organism and for being in turn protected against others, who, in virtue of that protection, are in their turn deprived of what they might call their liberty.

No one can have failed to observe that this interference with personal liberty becomes greater day by day. It is a tendency of modern governments, based presumably upon increased experience, to increase these protective regulations. Thus we have laws against adulteration of food, against the placing of buildings concerned with obnoxious trades in positions where people will be inconvenienced by them. We make persons suffering from infectious diseases isolate themselves, and if they cannot do this at home, we make them go to the fever hospital. Further, we insist upon the doctor, whose position resembles that of a confessor, breaking his obligation of professional secrecy and informing the authorities as to the illness of his patient. We interfere with the liberty of men and women to work as long as they like or to make their children labour for excessive hours. We insist upon dangerous machinery being fenced in. In a thousand ways we—the State—interfere with the liberty of our fellows. Finally, when the needs of the community are most pressing we interfere [Pg 82] most with the freedom of the subject. Thus, in these islands, we were recently living under a Defence of the Realm Act—with which no reasonable person quarrelled. Yet it forbad many things not only harmless in themselves but habitually permitted in times of peace. We were subject to penalties if we showed lighted windows: they must be shuttered or provided with heavy curtains. We might not travel in railway carriages at night with the blinds undrawn. The papers might not publish, nor we say in public, things which in time of peace would go unnoticed. There were a host of other matters to which allusion need not be made. Enough has been said to show that the State has and exerts the right to control the actions of those who belong to it, and that in time of stress it can and does very greatly intensify that control and does so without arousing any real or widespread discontent. Of course we all grumble, but then everybody, except its own members, always does more or less grumble at anything done by any government: that is the ordinary state of affairs. But at any rate we submit ourselves, more or less gracefully, to this restraint because we persuade ourselves or are persuaded that it is for the good of the State and thus for the good of ourselves, both as private individuals and as members of the State.

And many of us, at any rate, comfort ourselves with the thought that a great many of the regulations which appear to be most tyrannical and most to interfere with the natural liberty of [Pg 83] mankind are devised not with that end in view but with the righteous intention of protecting those weaker members of the body who are unable to protect themselves. If the State does not stand by such members and offer itself as their shield and support, it has no claim to our obedience, no real right to exist, and so we put up with the inconvenience, should such arise, on account of the protection given to the weaker members and often extended to those who would by no means feel pleased if they heard themselves thus described.

Let us substitute the Church for the State and let us remember that there are times when she is at closer grips with the powers of evil than may be the case at other times. The parallel is surely sufficiently close.

So far as earthly laws can control one, no one is obliged to be a member of the Catholic Church nor a citizen of the British Empire. I can, if I choose, emigrate to America, in process of time naturalise myself there and join the Christian Science organisation or any other body to which I find myself attracted. But as long as I remain a Catholic and a British citizen I must submit myself to the restrictions imposed by the bodies with which I have elected to connect myself. We arrive at the conclusion then that the ordinary citizen, even if he never adverts to the fact, is in reality controlled and his liberty limited in all sorts of directions.

Now the scientific man, in his own work, is subject to all sorts of limitations, apart altogether [Pg 84] from the limitations to which, as an ordinary member of the State, he has to submit himself.

He is restricted by science: he is not completely free but is bound by knowledge—the knowledge which he or others have acquired.

To say he is limited by it is not to say that he is imprisoned by it or in bondage to it. "One does not lose one's intellectual liberty when one learns mathematics," says the late Monsignor Benson in one of his letters, "though one certainly loses the liberty of doing sums wrong or doing them by laborious methods!"

Before setting out upon any research, the careful man of science sets himself to study "the literature of the subject" as he calls it. He delves into all sorts of out-of-the-way periodicals to ascertain what such a man has written upon such a point. All this he does in order that he may avoid doing a piece of work over again unnecessarily: *unnecessarily*, for it maybe actually necessary to repeat it, if it is of very great importance and if it has not been repeated and verified by other observers. Further, he delves into this literature because it is thus that he hopes to avoid the many blind alleys which branch off from every path of research, delude their explorer with vain hopes and finally bring him face to face with a blank wall. In a word the inquirer consults his authorities and when he finds them worthy of reliance, he limits his freedom by paying attention to them. He does not say: "How am I held in bondage by this assertion that the earth goes round the sun," but accepting that fact, he rejects such of his [Pg 85] conclusions as are obviously irreconcilable with it. Surely this is plain common sense and the man who acted otherwise would be setting himself a quite impossible task. It is the weakness of the "heuristic method" that it sets its pupils to find out things which many abler men have spent years in investigating. The man who sets out to make a research, without first ascertaining what others have done in that direction, proposes to accumulate in himself the abilities and the life-work of all previous generations of labourers in that corner of the scientific vineyard.

There is a somewhat amusing and certainly interesting instance of this which will bear quotation. The late Mr. Grant Allen, who knew something of quite a number of subjects though perhaps not very much about any of them, devoted most of his time and energies (outside his stories, some of which are quite entertaining) to not always very accurate essays in natural history. One day, however, his evil genius prompted him to write and, worse still, to publish a book entitled *Force and Energy: A Theory of Dynamics*, in which he purported to deal with a matter of which he knew far less even than he did about animated nature. Mark the inevitable result! A copy of the book was forwarded to the journal *Nature*, and sent by its editor to be dealt with by the competent hands of Sir Oliver (then Professor) Lodge. [27]

This is how that eminent authority dealt with [Pg 86] it. "There exists a certain class of mind," he commences, "allied perhaps to the Greek sophist variety, to which ignorance of a subject offers no sufficient obstacle to the composition of a treatise upon it." It may be rash to suggest that this type of mind is well developed in philosophers of the Spencerian school, though it would be possible to adduce some evidence in support of such a suggestion. "In the volume before us," he continues, "Mr. Grant Allen sets to work to reconstruct the fundamental science of dynamics, an edifice which, since the time of Galileo and Newton, has been standing on what has seemed a fairly secure and substantial basis, but which he seems to think it is now time to demolish in order to make room for a newly excogitated theory. The attempt is audacious and the result—what might have been expected. The performance lends itself indeed to the most scathing criticism; blunders and misstatements abound on nearly every page, and the whole thing is simply an emanation of mental fog." It would occupy too much space to reproduce this criticism with any fullness, but one or two points exceedingly germane to our subject can hardly go without notice. Alluding to a certain question, which seems to have greatly bothered Mr. Allen and likewise Mr. Clodd, who, it would appear, was associated with him in this performance, the reviewer says: "The puzzle was solved completely long ago, in the clearest possible manner, and the '*Principia*' is the witness to it; but it is still felt to be a difficulty by be [Pg 87] ginners, and I suppose there is no offence in applying this harmless epithet to both Mr. Grant Allen and Mr. Clodd, so far as the truths of dynamics and physics are concerned." One last quotation: "The thing which strikes one most forcibly about the physics of these paper philosophers is the extraordinary contempt which, if they are consistent, they must or ought to feel for men of science. If Newton, Lagrange, Gauss, and Thompson, to say nothing of smaller men, have muddled away their brains in concocting a scheme of dynamics wherein the very definitions are all wrong; if they have arrived at a law of conservation of energy without knowing what the word energy means, or how to define it; if they have to be set right by an amateur who has devoted a few weeks or months to the subject and acquired a rude smattering of some of its terms, 'what intolerable fools they must all be!'" Such is the result of asserting one's freedom by escaping the limitations of knowledge! We see what happens when a person sets out to deal with science untrammelled by any considerations as to what others have thought and established. The necessary result is that he plunges headforemost into all or most of the errors which were pitfalls to the first labourers in the field. Or, again, he painfully and uselessly pursues the blind alleys which they had wandered in, and from which a perusal of their works would have warned off later comers.

Oh, irony of fate! the same thing precisely happens when men of scientific eminence indulge [Pg 88] in religious dissertations, for of course, though it is not quite so obvious to such writers, the same blunder is quite possible in non-scientific fields of knowledge. I once asked one versed in theology what he thought of the religious articles of a distinguished man, unfamiliar himself with theology, yet, none the less, then splashing freely and to the great admiration of the ignorant, in the theological pool. His reply was that in so far as they were at all constructive, they consisted mostly of exploded heresies of the first century. Is not this precisely what one would have expected *a priori*? A man commencing to write on science or religion who neglects the work of earlier writers places himself in the position of the first students of the subject and very naturally will make the same mistakes as they made. He refuses to be hampered and biased by knowledge, and the result follows quite inevitably. "A scientist," says Monsignor Benson, "is hampered and biased by knowing the earth goes round the sun." The fact of the matter is that the man of science is not a solitary figure, a *chimæra bombinans in vacuo*. In whatever direction he looks he is faced by the figures of other workers and he is limited and "hampered" by their work. Nor are these workers all of them in his own area of country, for the biologist, for example, cannot afford to neglect the doings of the chemist; if he does he is bound to find himself led into mistakes. No doubt the scientific man is at times needlessly hampered by theories which he and others at the time take to be fairly [Pg 89] well established facts, but which after all turn out to be nothing of the kind. This in no way weakens the argument, but rather by giving an additional reason for caution, strengthens it.

If we carefully consider the matter we shall be unable to come to any other conclusion than that every writer, even of the wildest form of fiction, is in some way and to some extent hampered and limited by knowledge, by facts, by things as they are or as they appear to be. That will be admitted; but it will be urged that the hampering and limiting with which we have been dealing is not merely legitimate but inevitable, whereas the hampering and limiting—should such there be—on the part of the Church is wholly illegitimate and indefensible.

"All that you say is no doubt true," our antagonist will urge, "but you have still to show that your Church has any right or title to interfere in these matters. And even if you can make some sort of case for her interference, you have still to disprove what so many people believe, namely, that the right, real or assumed, has not been arbitrarily used to the damage, or at least to the delay of scientific progress. Chemistry," we may suppose our antagonist continuing, "no doubt has a legitimate right to have its say, even to interfere and that imperatively, where chemical considerations invade the field of biology, for example. But what similar right does religion possess? For instance," he might proceed, "some few years ago a distinguished physiologist, then occupying the Chair of the British Associa [Pg 90] tion, invoked the behaviour of certain chemical substances known as colloids in favour of his anti-vitalistic conclusions. At once he was answered by a number of equally eminent chemists that the attitude he had adopted was quite incompatible with facts as known to them; in a word, that chemistry disagreed with his ideas as to colloids. Everybody admitted that the chemists must have the final word on this subject: are you now claiming that religion or theology, or whatever you choose to call it, is also entitled to a say in a matter of that kind?" This supposititious conversation illustrates the confusion which exists in many minds as to the point at issue. One science is entitled to contradict another, just as one scientific man is entitled to contradict another on a question of fact. But on a question of *fact* a theologian is not entitled—*quâ* theologian—nor would he be expected to claim to be entitled, to contradict a man of science.

It ought to be widely known, though it is not, that the idea that theologians can or wish to intrude—again *quâ* theologians—in scientific disputes as to chemical, biological, or other facts, is a fantastic idea without real foundation save that of the one mistake of the kind made in the case of Galileo and never repeated—a mistake, let us hasten to add, made by a disciplinary authority and—as all parties admit—in no way involving questions of infallibility. To this case we will revert shortly. Meanwhile it may be briefly stated that the claim made by the Church is in connection with some few—some very few—of [Pg 91] the *theories* which men of science build up upon the facts which they have brought to light. Some of these theories do appear to contradict theological dogmas, or at least may seem to simple people to be incompatible with such dogmas, just as the people of his time—Protestants by the way, no less than Catholics—did really think that Galileo's theory conflicted with Holy Writ. In such cases, and in such cases alone, the Church holds that she has at least the right to say that such a theory should not be proclaimed to be true until there is sufficient proof for it to satisfy the scientific world that the point has been demonstrated.

This is really what is meant by the tyranny of the Church; and it may now be useful to consider briefly what can be said for her position. We must begin by looking at the matter from the Church's standpoint. It is a good rule to endeavour to understand your opponent's position before you try to confute him; an excellent rule seldom complied with by anti-Catholic controversialists. Now the Church starts with the proposition that man has an immortal soul destined to eternal happiness or eternal misery, and she proceeds to claim that she has been divinely constituted to help man to enjoy a future of happiness. Of course these are opinions which all do not share, and with the arguments for and against which we cannot here deal. If a man is quite sure that he has no soul and that there is no hereafter there is nothing more to be said than: "Let us eat and drink, for to-morrow [Pg 92] we die." Nothing very much matters in this world except that we should make ourselves as comfortable as we can during the few years we have to spend in it.

Again, there are others who, whilst believing the first doctrine set down above, will have none of the other. With them we enter into no argument here, and only say that to have a guide is better than to have no guide. Catholics, who accept gratefully her guidance, do believe that the Church can help a man to save his soul, and that she is entrusted, to that end, with certain powers. Her duty is to preserve and guard the Christian Revelation—the scheme of doctrine regarding belief and conduct by which Jesus Christ taught that souls were to be saved. She is not an arbitrary ruler. Her office is primarily that of Judge and Interpreter of the deposit of doctrine entrusted to her.

In this she claims to be safeguarded against error, though her infallible utterances would seem incredibly few, if summed up and presented to the more ignorant of her critics. She also claims to derive from her Founder legislative power by which she can make decrees, unmake them or modify and vary them to suit different times and circumstances. She rightfully claims the obedience of her children to this exercise of her authority, but such disciplinary enactments, by their very nature variable and modifiable, do not and cannot come within the province of her infallibility, and admittedly they need not be always perfectly wise or judicious. Such dis [Pg 93] ciplinary utterances, it may be added, at least in the field of which we are treating, indeed in any field, are also incredibly few when due regard is had to the enormous number of cases passing under the Church's observation.

We saw just now that the State exercised a very large jurisdiction for the purpose of protecting the weak who were unable or little able to protect themselves. It is really important to remember, when we are considering the powers of the Church and her exercise of them, that these disciplinary powers are put in operation, not from mere arrogance or an arbitrary love of domination—as too many suppose—but with the primary intention of protecting and helping the weaker members of the flock. If the Church consisted entirely of theological experts a good deal of this exercise of disciplinary power might very likely be regarded as wholly unnecessary. Thus the Church freely concedes not only to priests and theologians, but to other persons adequately instructed in her teaching, full permission to read books which she has placed on her black list or *Index*—from which, in other words, she has warned off the weaker members of the flock.

The net of Peter, however, as all very well know, contains a very great variety of fish, and—to vary the metaphor—to the fisherman was given charge not only of the sheep—foolish enough, heaven knows!—but also of the still more helpless lambs. Thus it becomes the duty and the privilege of the successors of the fisherman to protect the sheep and the lambs, and not merely [Pg 94] to protect them from wild beasts who may try to do harm from without, but quite as much from the wild rams of the flock who are capable of doing a great deal of injury from within. In one of his letters, from which quotation has already been made, the late Monsignor Benson sums up, in homely, but vivid language, the point with which we have just been dealing. "Here are the lambs of Christ's flock," he writes: "Is a stout old ram to upset and confuse them when he needn't ... even though he is right? The flock must be led gently and turned in a great curve. We can't all whip round in an instant. We are tired and discouraged and some of us are exceedingly stupid and obstinate. Very well; then the rams can't be allowed to make brilliant excursions in all directions and upset us all. We shall get there some day, if we are treated patiently. We are Christ's lambs after all."

The protection of the weak: surely, if it be deemed both just and wise on the part of the civil government to protect its subjects by legislation in regard to adulterated goods, contagious diseases, unhealthy workshops and dangerous machinery, why may not the Church safeguard her children, especially her weaker children, the special object of her care and solicitude, from noxious intellectual foods?

It is just here that the question of the *Index* arises. Put briefly, this is a list of books which are not to be read by Catholics unless they have permission to read them—a permission which, as we have just seen, is never refused when any [Pg 95] good reason can be given for the request. I can understand the kind of person who says: "Exactly, locking up the truth; why not let everybody read just what they like?" To which I would reply that every careful parent has an *Index Prohibitorius* for his household; or ought to have one if he has not. I once knew a woman who allowed her daughter to plunge into *Nana* and other works of that character as soon as she could summon up enough knowledge of French to fathom their meaning. The daughter grew up and the result has not been encouraging to educationists thinking of proceeding on similar lines. The State also has its *Index Prohibitorius* and will not permit indecent books nor indecent pictures to be sold. Enough: let us again clear our minds of cant. There is a limit with regard to publications in every decent State and every decent house: it is only a question where the line is drawn. It is obvious that the Church must be permitted at least as much privilege in this matter as is claimed by every respectable father of a family.

We need not pursue the question of the *Index* any further, but before we leave it let us for a moment turn to another accusation levelled against Catholic men of science by anti-Catholic writers, that of concealing their real opinions on scientific matters, and even of professing views which they do not really hold, out of a craven fear of ecclesiastical denunciations. The attitude which permits of such an accusation is hardly courteous, but, stripped of its verbiage, that is the accusation as [Pg 96] it is made. Now, as there are usually at least some smouldering embers of fire where there is smoke, there is just one small item of truth behind all this pother. No Catholic, scientific man or otherwise, who really honours his Faith would desire wilfully to advance theories apparently hostile to its teaching. Further, even if he were convinced of the truth of facts which might appear—it could only be "appear"—to conflict with that teaching, he would, in expounding them, either show how they could be harmonised with his religion, or, if he were wise, would treat his facts from a severely scientific point of view and leave other considerations to the theologians trained in directions almost invariably unexplored by scientific men. Perhaps the memory of old, far-off, unhappy events should not be recalled, but it is pertinent to remark that the troubles in connection with a man whose name once stood for all that was stalwart in Catholicism, did not originate in, nor were they connected with, any of the scientific books and papers of which the late Professor Mivart was the author, but with those theological essays which all his friends must regret that he should ever have written.

It may not be waste of time briefly to consider two of the instances commonly brought up as examples when the allegation with which we are dealing is under consideration.

First of all let us consider the case of Gabriel Fallopius, who lived—it is very important to note the date—1523-1562; a Catholic and a churchman. Now it is gravely asserted that Fallopius [Pg 97] committed himself to misleading views, views which he knew to be misleading, because he thought that he was thereby serving the interest of the Church. What he said concerned fossils, then beginning to puzzle the scientific world of the day. Confronted with these objects and living, as he did, in an unscientific age, when the seven days of creation were interpreted as periods of twenty-four hours each and the universality of the Noachian deluge was accepted by everybody, it would have been something like a miracle if he had at once fathomed the true meaning of the shark's teeth, elephant's bones, and other fossil remains which came under his notice. His idea was that all these things were mere concretions "generated by fermentation in the spots where they were found," as he very quaintly and even absurdly put it. The accusation, however, is not that Fallopius made a mistake—as many another man has done—but that he deliberately expressed an opinion which he did not hold and did so from religious motives. Of course, this includes the idea that he knew what the real explanation was, for had he not known it, he could not have been guilty of making a false statement. There is no evidence whatever that Fallopius ever had so much as a suspicion of the real explanation, nor, it may be added, had any other man of science for the century which followed his death.

Then there arose another Catholic churchman, Nicolaus Stensen (1631-1686), who, by the way, ended his days as a bishop, who did solve the [Pg 98] riddle, giving the answer which we accept to-day as correct, and on whom was conferred by his brethren two hundred years later the title of "The Father of Geology." It is a little difficult to understand how the "unchanging Church" should have welcomed, or at least in no way objected to, Stensen's views when the mere entertainment of them by Fallopius is supposed to have terrified him into silence. But when the story of Fallopius is mistold, as indicated above, it need hardly be said that the story of Stensen is never so much as alluded to.

The real facts of the case are these: Fallopius was one of the most distinguished men of science of his day. Every medical student becomes acquainted with his name because it is attached to two parts of the human body which he first described. He made a mistake about fossils, and that is the plain truth—as we now know, a most absurd mistake, but that is all. As we hinted above, he is very far from being the only scientific man who has made a mistake. Huxley had a very bad fall over *Bathybius* and was man enough to admit that he was wrong. Curiously enough, what Huxley thought a living thing really was a concretion, just as what Fallopius thought a concretion had been a living thing.

Another extremely curious fact is that another distinguished man of science, who lived three hundred years later than Fallopius and had all the knowledge which had accumulated during that prolific period to assist him, the late Philip Gosse, fell into the same pit as Fallopius. As his son [Pg 99] tells us, he wrote a book to prove that when the sudden act of creation took place the world came into existence so constructed as to bear the appearance of a place which had for æons been inhabited by living things, or, as some of his critics unkindly put it, "that God hid the fossils in the rocks in order to tempt geologists into infidelity." Gosse had the real answer under his eyes which Fallopius had not, for the riddle was unread in the latter's days. Yet Gosse's really unpardonable mistake was attributed to himself alone, and "Plymouth Brethrenism," which was the sect to which he belonged, was not saddled with it, nor have the Brethren been called obscurantists because of it.

Of course there is a second string to the accusation we are dealing with. If the scientific man did really express new and perhaps startling opinions, they would have been much newer and much more startling had he not held himself in for fear of the Church and said only about half of what he might have said. It is the half instead of the whole loaf of the former accusation. Thus, in its notice of Stensen, the current issue of the *Encyclopædia Britannica* says: "Cautiously at first, for fear of offending orthodox opinion, but afterwards more boldly, he proclaimed his opinion that these objects (*viz.* fossils) had once been parts of living animals."

One may feel quite certain that if Stensen had not been a Catholic ecclesiastic this notice would have run—and far more truthfully— "Cautiously at first, until he felt that the facts at his disposal [Pg 100] made his position quite secure, and then more boldly, etc. etc."

What in the ordinary man of science is caution, becomes cowardice in the Catholic. We shall find another example of this in the case of Buffon (1707-1788) often cited as that of a man who believed all that Darwin believed and one hundred years before Darwin, and who yet was afraid to say it because of the Church to which he belonged. This mistake is partly due to that lamentable ignorance of Catholic teaching, not to say that lamentable incapacity for clear thinking, on these matters, which afflicts some non-Catholic writers. Let us take an example from an eminently fairly written book, in which, dealing with Buffon, the author says: "I cannot agree with those who think that Buffon was an out-and-out evolutionist, who concealed his opinions for fear of the Church. No doubt he did trim his sails—the palpably insincere *Mais non, il est certain par la révélation que tous les animaux ont également participé à la grâce de la création*, following hard upon the too bold hypothesis of the origin of all species from a single one, is proof of it." Of course it is nothing of the kind, for, whatever Buffon may have meant, and none but himself could tell us, it is perfectly clear that whether creation was mediate (as under transformism considered from a Christian point of view it would be) or immediate, every created thing would participate in the grace of creation, which is just the point which the writer from whom the quotation has been made has missed.

[Pg 101]

The same writer furnishes us with the real explanation of Buffon's attitude when he says that Buffon was "too sane and matter-of-fact a thinker to go much beyond his facts, and his evolution doctrine remained always tentative." Buffon, like many another man, from St. Augustine down to his own times, considered the transformist explanation of living nature. He saw that it unified and simplified the conceptions of species and that there were certain facts which seemed strongly to support it. But he does not seem to have thought that they were sufficient to establish it and he puts forward his views in the tentative manner which has just been suggested.

The fact is that those who father the accusations with which we have been dealing either do not know, or scrupulously conceal their knowledge, that what they proclaim to be scientific cowardice is really scientific caution, a thing to be lauded and not to be decried.

Let us turn to apply the considerations with which we have been concerned to the case of Galileo, to which generally misunderstood affair we must very briefly allude, since it is the standby of anti-Catholic controversialists. Monsignor Benson, in connection with the quotation recently cited, proclaimed himself "a violent defender of the Cardinals against Galileo." Perhaps no one will be surprised at his attitude, but those who are not familiar with his *Life and Letters* will certainly be surprised to learn that Huxley, after examining into the question, "arrived at the conclusion that the Pope and [Pg 102] the College of Cardinals had rather the best of it." [28]

None the less it is the stock argument. Father Hull, S. J., whose admirable, outspoken, and impartial study of the case [29] should be on everybody's bookshelves, freely admits that the Roman Congregations made a mistake in this matter and thus takes up a less favourable position towards them than even the violently anti-Catholic Huxley.

No one will deny that the action of the Congregation was due to a desire to prevent simple persons from having their faith upset by a theory which seemed at the time to contradict the teaching of the Bible. Remember that it was only a theory and that, when it was put forward, and indeed for many years afterwards, it was not only a theory, but one supported by no sufficient evidence. It was not in fact until many years after Galileo's death that final and convincing evidence as to the accuracy of his views was laid before the scientific world. There can be but little doubt that if Galileo had been content to discuss his theory with other men of science, and not to lay it down as a matter of proved fact—which, as we have seen, it was not—he would never have been condemned. Whilst we may admit, with Father Hull, that a mistake was made in this case, we may urge, with Cardinal Newman, that it is the only case in which such a thing has happened—surely a remarkable fact. It is not [Pg 103] for want of opportunities. Father Hull very properly cites various cases where a like difficulty might possibly have arisen, but where, as a matter of fact, it has not. For example, the geographical universality of the Deluge was at one time, and that not so very long ago, believed to be asserted by the Bible; while, on the other hand, geologists seemed to be able to show, and in the event did show, that such a view was scientifically untenable. The attention of theologians having been called to this matter, and a further study made of passages which until then had probably attracted but little notice, and quite certainly had never been considered from the new point of view, it became obvious that the meaning which had been attached to the passages in question was not the necessary meaning, but on the contrary, a strained interpretation of the words. No public fuss having arisen about this particular difficulty, the whole matter was gradually and quietly disposed of. As Father Hull says, "the new view gradually filtered down from learned circles to the man in the street, so that nowadays the partiality of the Deluge is a matter of commonplace knowledge among all educated Christians, and is even taught to the rising generation in elementary schools."

In accordance with the wise provisions of the Encyclical *Providentissimus Deus*, with which all educated Catholics should make themselves familiar, conflicts have been avoided on this, and on other points, such as the general theory of evolution and the various problems connected [Pg 104] with it; the antiquity of man upon the earth and other matters as to which science is still uncertain. Some of these points might seem to conflict with the Bible and the teachings of the Church. As Catholics we can rest assured that the true explanation, whenever it emerges, cannot be opposed to the considered teaching of the Church. What the Church does—and surely it must be clear that from her standpoint she could not do less—is to instruct Catholic men of science not to proclaim *as proved facts* such modern theories—and there are many of them—as still remain wholly unproved, when these theories are such as might seem to conflict with the teaching of the Church. This is very far from saying that Catholics are forbidden to study such theories.

On the contrary, they are encouraged to do so, and that, need it be said, with the one idea of ascertaining the truth? Men of science,

Catholic and otherwise, have, as a mere matter of fact, been time and again encouraged by Popes and other ecclesiastical authorities to go on searching for the truth, never, however, neglecting the wise maxim that all things must be proved. So long as a theory is unproved, it must be candidly admitted that it is a crime against science to proclaim it to be incontrovertible truth, yet this crime is being committed every day. It is really against it that the *magisterium* of the Church is exercised. The wholesome discipline which she exercises might also be exercised to the great benefit of the ordinary reading public by some [Pg 105] central scientific authority, can such be imagined, endowed with the right to say (and in any way likely to be listened to): "Such and such a statement is interesting—even extremely interesting—but so far one must admit that no sufficient proof is forthcoming to establish it as a fact: it ought not, therefore, to be spoken of as other than a theory, nor proclaimed as fact."

Such constraint when rightly regarded is not or would not be a shackling of the human intellect, but a kindly and intelligent guidance of those unable to form a proper conclusion themselves. Such is the idea of the Church in the matter with which we have been dealing.

FOOTNOTES:

[Pg 106]

[23] *Darwiniana*, p. 147.

[24] See, for example, his *Life and Letters*, i., 307.

[25] *Hume, English Men of Letters Series*, p. 135.

[26] Of course, it may be argued, no Fellow need have applied for an *imprimatur*; he did it *ex majori cantelâ* as the lawyers say. This may be so, but the same applies to the ecclesiastical *imprimatur*.

[27] The review from which the following quotations are made appeared in *Nature* on January 24, 1889.

[28] Vol. ii., p. 113.

[29] *Galileo and His Condemnation*, Catholic Truth Society of England.

V. SCIENCE AND THE WAR

Amongst various important matters now brought to a sharper focus in the public eye, few, if any, require more careful attention than that which is concerned with science, its value, its position, its teachings, and how it should be taught. No one who has followed the domestic difficulties due to our neglect of the warnings of scientific men can fail to see how we have had to suffer because of the lax conduct of those responsible for these things in the past.

Within the first few weeks after the war broke out—to take one example—every medical man was the recipient of a document telling him of the expected shortage in a number of important drugs and suggesting the substitutes which he might employ. It was a timely warning; but it need never have been issued if we had not allowed the manufacture of drugs, and especially those of the so-called "synthetic" group, to drift almost entirely into the hands of the Badische Aniline Fabrik, and kindred firms in Germany. This difficulty, now partly overcome, is one which never would have arisen but for the deaf ear turned to the warnings of the scientific chemists. [Pg 107] British pharmaceutical chemists, with one or two exceptions, had been relying upon foreign sources not only for synthetic drugs but actually for the raw materials of many of their preparations—such, for example, as aconite, belladonna, henbane, all of which can be freely grown—which even grow wild—in these islands; even, incredible as it may seem, for foxglove leaves. These things with many others were imported from Germany and Austria. Here again leeway has had to be made up; but it ought never to have been necessary, and now that the war is over steps should be taken to see that it never need be necessary again. The encouragement of British herb-gardens and of scientific experiment therein on the best method of culture for the raw material of our organic medicines must certainly be matters early taken in hand.

The classical example of the mortal injury done to British manufacture by the British manufacturer's former contempt for the scientific man is that of the aniline dyes, which are so closely associated with the synthetic drugs as to form one subject of discussion. Quite early in the war dye-stuffs ran short, and there was no means of replenishing the stock in Britain, nor even in America, these products having formed the staple of a colossal manufacture, with an enormous financial turnover, in Germany.

Let us look at the history of these dyes. The first aniline dye was discovered quite by accident, in 1856, by the late Professor W. H. Perkin. He called it "mauve," from the French word for the [Pg 108] mallow, the colour of whose flower it somewhat resembled. In 1862 there was an International Exhibition in London; and those who remembered it and its predecessor of 1851 have declared that the case of aniline dye-stuffs—for by that time quite a number of new pigments had been discovered—excited at the later the same attention as that given to the Koh-i-noor at the earlier. The invention, out of which grew the enormous German business already alluded to, and with which has been associated the discovery and manufacture of the synthetic drugs, was entirely British in its inception and in its early stages. Moreover the raw materials on which it depended, namely, gas-tar products, were to be had in greater abundance in England than anywhere else. Yet, at the time when the war broke out, this industry had been allowed almost entirely to drift into German hands.

How was this? Let an expert reply. It was due, he tells us, to the neglect of "the repeated warnings which have been issued since that time" (*viz.* 1880, by which date the Germans had succeeded in capturing the trade in question) "in no uncertain voice by Meldola, Green, the Perkins (father and son), and many other English chemists." Further, he continues, two causes have invariably been indicated for the transfer of this industry to Germany—"first the neglect of organic chemistry in the Universities and colleges of this country" (a neglect which has long ceased), "and then the disregard by manufacturers of scientific methods and assistance and total in [Pg 109] difference to the practice of research in connection with their processes and products." I remember talking some twenty-five years ago to a highly educated young student of Birmingham who was of German parentage though of English birth. He had just taken the degree of Doctor of Science in London University, and was on the eve of abandoning the adopted country of his parents for a position in the research laboratories of the Badische company, where he would be one among a number of chemists, running into hundreds, all engaged in research on gas-tar products. At that moment the great Birmingham gas-company was employing the services of one trained chemist.

Such was and is the neglect of science by business men. Could it have been otherwise, considering their bringing up? Let me again be reminiscent. I suppose the public school in England (not a Catholic school, for I was then a Protestant) at which I pursued what were described as studies did not in any very marked degree differ from its sister schools throughout the country. How was science encouraged there? One hour per week, exactly one-fifth of the time devoted weekly, not to Greek and Latin (that would have been almost sacrilegious), but to the writing of Greek and Latin prose and alleged Greek and Latin verse—that was the amount of time which was devoted to what was called science. I suppose I had an ingrained vocation for science, for it was the only subject, except English composition, in which I ever felt interest at school. If the vocation had not been there, [Pg 110] any interest in the subject must necessarily have been slain once for all in me, as I am sure it was in scores of others, by the way it was taught; for the instruction was confided to the ordinary form-master, who doled out his questions from a text-book perfunctorily used and probably heartily despised by a man brought up on strict classical or mathematical lines. Our manufacturer is brought up in a school of this kind, and it would be a miracle if he emerged from it with any respect for science. Things have changed now, and for the better, as they have at most of the Universities; but we are dealing with the generation of manufacturers of my age who were largely responsible for the neglects now in question. Well, the boy left his school and went to Oxford or Cambridge, neither of which then greatly encouraged science. Its followers were, I believe, known as "Stinks Men." At any rate it is only comparatively recently that we have seen the splendid developments of to-day in those ancient institutions. One relic of the ancient days gives us an illuminating idea of how things used to be, just as a fossil shows us the environment of its day. [30] Trinity College, Dublin, has fine provision for scientific teaching, and a highly competent staff to teach. But in its constitution it shows the attitude towards science which till lately informed the older Universities.

[Pg 111]

Trinity College has in its Fellowship system one of the most important series of pecuniary rewards perhaps in Europe, of an educational character. A man has only once to pass an examination, admittedly one of great severity and competitive in character, and thenceforward to go on living respectably and doing such duties as are committed to him, to be ensured an excellent and increasing income for life. How great the rewards are will be gathered from the fact that a distinguished occupant of one of these positions some years ago endeavoured—with complete success—to enforce on me the importance of the Fellowship examination by telling me that he had already received over £50,000 in emoluments as a result of his success. He has received a good deal more since, and I hope will continue to be the recipient of this shower of gold for many years to come. [31] No doubt much might be urged for this system, which was for a long time popular in China for the selection of Mandarins, and I am not criticising it here. What I want to emphasise is that the examination for these valuable positions is either classical or mathematical, and there it ends. The greatest biologist in the world would have as much chance of a Fellowship as the ragged urchin in the street unless he could "settle Hoti's business" or elucidate [Greek: P] or do other things of that kind. It is a luminous example of what was—must we say [Pg 112] is?—thought of science in certain academic circles. Of course it may be urged—I have actually heard it urged—that nothing is science save that which is treatable by mathematical methods. It was a kind of inverted M. Jourdain who used this argument, a gentleman who imagined himself to have been teaching science during a long life without ever having effected what he supposed to be his object. Then, again, our manufacturer, whose object in life is to make money, is naturally, perhaps even necessarily, affected by the kind of salaries which highly trained and highly eminent men of science receive by way of reward for their work. Few, if any, receive anything like the emoluments attaching to the position of County Court Judge, and I know of only one case in which a Professor's income, to the delight and envy of all the teaching profession, actually, for a few years, soared somewhat near the empyrean of a Puisne Judge's reward.

Perhaps this is not to be wondered at; for Parliament always contains many lawyers, and at the moment, I think, not a single scientific expert, at least among the Commons. This is not really a sordid argument, though it may appear so. The labourer, after all, is worthy of his hire; but in the scientific world it very, very seldom happens that the hire is worthy of the labourer. Even to this day there is plenty of truth in the description of the attitude of Mr. Meagles towards Mr. Doyce as detailed by the author of *Little Dorrit*. Perhaps that is partly because it is generally the [Pg 113] man of business, and not the unhappy man of science, who gains the money produced by scientific discoveries. These are often, if not usually, made by accident, and by a man on the track of something else, on the elucidation of which he is probably so intent that he cannot spare time for side-issues, very likely never even thinks of them. Sir James Dewar discovered the principle of the "Thermos flask" whilst he was working at the exceedingly difficult subject of the liquefaction of air. I hope Sir James had the prescience to patent his discovery, and reap the reward which was due to him; but, if he did, he is one amongst a thousand who never took this trouble and of whom *Sic vos non vobis* might well be said. When Sabatier had shown the importance of combinations of hydrogen effected by what is known as a catalyst, numerous patents were taken out—by other people, of course—on which were founded very flourishing businesses. Sabatier profited by none of these—so I understand. He received a Nobel prize for his discoveries; but another hath his heritage.

Though science has not received any great encouragement, yet in spite of that—the cynic might say because of that—it has made amazing progress during the past half-century. Mr. Chesterton somewhere notes that "a time may easily come when we shall see the great outburst of science in the Nineteenth Century as something quite as splendid, brief, unique, and ultimately abandoned as the outburst of art at the Renaissance." That, of course, may be so, but as to [Pg 114] the outburst there can be no question, nor of its persistence to the present day. That also is surely a curious phenomenon; for, as regards most other things, we seem to be in the trough of the wave, and not merely in these islands but all over the civilised world. In Art, in Music, in Literature, in the Drama, it would be difficult to argue in favour of a pre-eminence, or even of an equality of the present age, comparing it with its predecessors.

Take the politicians of the world; it is perhaps difficult, even foolish, for us who are living with them to prophesy with any approximation of accuracy what the historian of a future day may say about them. He may sum them up as respectable, honest mediocrities trying to do their best under exceptionally difficult circumstances; he may put them lower; he may put them higher; he may differentiate between those of different nations; but there is little doubt that, with the exception of the American President, he will not be able to point to any one of the calibre of Pitt or of Bismarck or of the less severely tried Disraeli or Gladstone.

But just the reverse is the case in science, which has men of the very first rank living, working, and discovering to-day. There are indeed signs that even our Government is cognizant of this. The creation of a Department of Industrial Scientific Research, the provision of a substantial income for the same, the increase of research-grants to learned societies, these and other things show that some attempt will be made to recognise the [Pg 115] value of science to the State. Further, the lesson seems to have gone home to some few at least that there is no difference between what have been absurdly called Pure and Applied Science, since so very many "Applied" discoveries—such as the "Thermos"—arose in the course of what certainly would have been described as "Pure" researches.

It is to the public advantage that every educated person should know something about science; nor is this by any means as big or difficult an achievement as some may imagine. It is not necessary to teach any very large number of persons very much about any particular science or group of sciences. What is really important is that people should imbibe some knowledge of scientific methods—of the meaning of science. This can be done from the study of quite a few fundamental propositions of any one science under a good teacher—a first essential. Any person thus educated will, for the remainder of his life, be able at least to understand what is meant by science and the scientific method of approaching a problem. He will not, like an educational troglodyte at a recent Conference, refuse to describe anything as science which is not capable of mathematical treatment, nor allude compendiously to physiological study as "the cutting up of frogs." In a word, he will be an educated man, which can no more be said of one ignorant of science than it can be of one whose mind has never experienced the softening influence of letters.

[Pg 116]

So far, everybody whose opinion counts seems to be agreed; but in any plea for an extended and improved teaching of science, certain points ought not to be left out of count. In the first place, science is not the key to all locks; there are many important things—some of the most important things in life—with which it has nothing whatever to do. It will be well to recall Mr. Balfour's words at the opening of the National Physical Laboratory: "Science depends on measurement, and things not measurable are therefore excluded, or tend to be excluded, from its attention. But Life and Beauty and Happiness are not measurable. If there could be a unit of happiness, politics might begin to be scientific." It follows that there are a number of subjects on which the scientific man is just as fit, or as unfit, to express an opinion as any other man. The intense preoccupation which serious scientific studies demand, may render the man who is engaged therein even less competent to express an opinion on alien subjects than one whose attention, less concentrated, has time to range over diverse fields of study. Readers of Darwin's *Life* will remember his confession that he had lost all taste for music, art, and literature; that he "could not endure to read a line of poetry" and found Shakespeare "so intolerably dull that it nauseated" him; and finally, that his mind seemed "to have become a kind of machine for grinding general laws out of a large collection of facts."

Despite this warning as to the limits of science, [Pg 117] we have no lack of instances of scientific men posing as authorities on subjects on which they had no real right to be heard, and, what is worse, being accepted as such by the uninstructed crowd. Thus Professor Huxley, who, as some one once said, "made science respectable," was wont to utter pontifical pronouncements on the subject of Home Rule for Ireland. His knowledge of that country was quite rudimentary, and his visits to it had been as few and as brief as if he had been its Sovereign; but that did not prevent him from delivering judgment, nor unfortunately deter many from following that judgment as if it had been inspired. I am not now arguing as to the rights and wrongs of Huxley's view on the matter in question: I have my own opinion on that. What I am urging is that his position, whether as a zoologist or, incidentally, as a great master of the English language, in no way entitled him to express an opinion or rendered him a better authority on such a question than any casual fellow-traveller in a railway carriage might easily be.

This is bad enough; but what is far worse is when scientific experts on the strength of their study of Nature assume the right of uttering judicial pronouncements on moral and sociological questions, judgments some at least of which are subversive of both decency and liberty. Thus we have lately been told that it is "wanton cruelty" to keep a weak or sickly child alive; and the medical man, under a reformed system of medical ethics, is to have leave and licence to [Pg 118] put an end to its life in a painless manner. To what enormities and dastardly agreements this might lead need hardly be suggested; and I am quite confident that the members of the honourable profession of physic, to which I am proud to belong, have no desire whatever for such a reform of the law or of their ethics. Then we are told in the same address (Bateson, *British Association Addresses in Australia*, 1914) that on the whole a decline in the birth-rate is rather a good thing, and that families averaging four children are quite enough to keep the world going comfortably. The date of this address will be noted; and the fact that the war, which was then just beginning, has probably caused its author and has caused everybody else to see the utter futility of such assertions.

However, if we are to rear only four children per marriage, and if we are to give the medical man liberty to weed out the weaklings, it behoves us to see that the children whom we produce are of the best quality. Let us, therefore, hie to the stud-farm, observe its methods and proceed to apply them to the human race. We must definitely prevent feeble-minded persons from propagating their species. Within limits, that is a proposition with which all instructed persons would agree, though few, we imagine, would put their opinions so uncharitably as the lecturer did: "The union of such social vermin we should no more permit than we would allow parasites to breed on our own bodies." But we must go farther than this, and introduce all sorts of [Pg 119] restrictions on matrimony, until finally it comes to be a matter to be arranged under rigid laws by a jury of elderly persons—all, we may feel perfectly sure, "cranks" of the first water.

In what *milieu* are their findings to take effect? It is very important to consider that. The author from whom I have been quoting tells us what we want to know. Man, he tells us, is "a rather long-lived animal, with great powers of enjoyment, if he does not deliberately forgo them." In the past, we are told, "superstitious and mythical ideas of sin have predominantly controlled these powers." We have changed all that now; as the parent in *Punch* says to the crying child by the seashore, "You've come out to enjoy yourself, and enjoy yourself you shall!" So we are to plunge into the whirlpool of eugenic delights without any fear of that "bugbear of a hell" which another writer congratulates us on getting rid of. We can, it appears, enter upon our eugenic experiment without a single moral scruple to restrain us or a single religious restriction to interfere with us. In this soil is the plant to be grown, and the first weed to be eradicated is that of the right of personal choice of a partner for life, or for such other term as the law under the new *régime* may require. Jack is to be torn from weeping Jill, and handed over to reluctant Joan, to whom he is personally displeasing and for whom he has not the slightest desire, and handed over because the Breeding Committee think it is likely to prove advantageous for the Coming Race. All that may be possible—or may not— [Pg 120] but what then? When you are carrying out Mendelian experiments on peas, you can enclose your flowers in muslin bags and prevent anything interfering with your observations. And in the stud-farm you can keep the occupants shut up.

But what are you going to do with Jack? and with Jill? And still more with Joan? They cannot be permanently isolated, neither are they restrained by any "mythical ideas of sin." They have been educated to the idea that their highest duty is to enjoy themselves. Why should they not do what they like? And consequently, as any reasoning person can see, "The Inevitable" must happen; and where is your experiment and where the Coming Race? It is perfectly useless for doctrinaires to argue, as doctrinaires will, about ethical restraints. Nature has *no* ethical restraints; and any ethical restraints which man has come from that higher nature of his which he does not share with the lower creation. What those whom the late Mr. Devas so aptly called "after-Christians" always forget is that the humane, the Christian side of life, which they as well as others exhibit, is due to the influence, lingering if you like, of Christianity. They ignore or forget the pit out of which they were digged.

By another Eugenist we are told that willy-nilly every sound, healthy person of either sex must get married or at least betake him or herself to the business of propagating the race. That at least is the essence of his singularly offensive dictum that since the celibacy of the Catholic clergy and of members of Religious [Pg 121] Orders deprives the State of a number of presumably excellent parents, "if monastic orders and institutions are to continue, they should be open only to the eugenically unfit." [32] If the religious call is not to be permitted to dispense a man or woman from entering the estate of matrimony, it may be assumed that nothing else, except an unfavourable report from the committee of selection, will do so. And, further, as the one object of all this is to bring super-children into the world, we must also assume that those who fail in this duty will find themselves in peril of the law.

Surely what has been set down shows that whatever scientific reputation the writers in question possess, and it is undeniably great, it has not equipped them, one will not merely say with moral or religious ideas, but with an ordinary knowledge of human nature. It has not equipped them with any conception apparently of political possibilities; and it has left them without any of that saving salt, a sense of humour. Like Huxley, they have started out to give opinions without first having made themselves familiar with the subject on which they were to deliver judgment.

It is perhaps little to be wondered at that the intense preoccupation which the study of science entails should tend to induce those whose attention is constantly fixed on Nature to imagine that from Nature can be drawn not only lessons of [Pg 122] physical life but lessons also of conduct. Of course this is quite wrong; for Nature has no moral lesson to teach us. We are told to go to the ant—at least the sluggard is—but for what? To amend his sluggardliness. No one has ever suggested that we should go to Nature to learn to be humble, kindly, unselfish, tolerant, and Christian, in our dealings with others; and for this excellent reason, that none of these things can be learnt from Nature. Science is neither moral nor immoral, but non-moral; and, as we have seen a thousand times in this present war, its kindest gifts to man can be used, and are used, for his cruel destruction. In this war, pre-eminently amongst all wars, we have the application of pure natural principles unameliorated by the influences of Christianity, or of chivalry, Christianity's offspring. As Sir Robert Borden has summed it up, German kultur is an attempt "to impose upon us the law of the jungle."

Natural Selection, some would have us believe, is the dominant law of living nature, and all would agree that it is an important law. Let us then, if we are to follow Nature, put it into practice. But Natural Selection means the Survival of the Fittest in the Struggle for Life. It consequently means the Extermination of the Less Fit, a little fact often left out of count. It means in three words "Might is Right," and was not that exactly the proposition by which we were confronted in this war? If Natural Selection be our only guide, let us sink hospital [Pg 123] ships, destroy innocent villages and towns, exterminate our weaker opponents in any way that seems best to us. It was all summed up centuries ago by the author of the Book of Wisdom: "Let us oppress the poor just man, and not spare the widow, nor honour the ancient grey hairs of the aged. But let your strength be the law of justice: for that which is feeble is found to be nothing worth." That is Natural Selection in operation in human life when human beings have been stripped of all "mythical ideas of Sin:" not a pretty picture nor a condition of affairs under which we should like long to exist. Some of the other resemblances are less dreadful, but none the less instructive. Let us take the matter of Mimicry. There is a form of protective mimicry whereby the living thing is like unto its surroundings, and thus escapes its enemy. We find it in warfare in the use of khaki dress, in white overalls in snow-time, in other such expedients. But there is also a form of Aggressive Mimicry in which a deadly thing makes itself look like something innocent, as the wolf tried to look in "Little Red Riding Hood." "The Germans were beginning their attack on Haumont. Their front-line skirmishers, to throw us into confusion, had donned caps which were a faint imitation of our own, and also provided themselves with Red Cross brassards" (*The Battle of Verdun*. H. Dugard). Not to be tedious on this point, which really does not require to be laboured, let me finish with one quotation from a vivid series of war-pictures. Boyd Cable is writing of men [Pg 124] in the trenches: "Civilised Man, in his latest art of war, has gone back to be taught one more simple lesson by the beast of the field and the birds of the air; the armed hosts are hushed and stilled by the passing air-machine, exactly as the finches and field-mice of hedgerow and ditch and field are frozen to stillness by the shadow of a hovering hawk, the beat of its passing wing."

No; an existence passed under conditions of this kind and as the normal state of affairs is not an existence to be contemplated with equanimity. We are anxious that science and scientific teaching should be assisted in every possible way. But let us be quite clear that while science has much to teach us and we much to learn from her, there are things as to which she has no message to the world. The Minor Prophets of science are never tired of advising theologians to keep their hands off science. The Major Prophets are too busy to occupy themselves with such polemics. But the theologian is abundantly in his right in saying to the scientific writer "Hands off morals!" for with morality science has nothing to do. Let us at any rate avoid that form of kultur which consists in bending Natural History to the teaching of conduct, uncorrected by any Christian injunctions to soften its barbarities.

FOOTNOTES:

[Pg 125]

[30] Since these lines were written, this state of affairs has come to an end and the first Fellow has been elected for his purely scientific attainments, in the person of the distinguished geologist, Professor Joly, F.R.S.

[31] It was the late distinguished Provost, Sir John Mahaffy, at whose instance the change in the Fellowship system was introduced.

[32] Conklyn, *Heredity and Environment in the Development of Men*. Princeton University Press, 1915.

VI. HEREDITY AND "ARRANGEMENT"

Some years ago, when I was delivering a lecture at the Cathedral Hall of Westminster, in the course of the questioning which took place at the termination of the discourse, which was on vitalism, I was asked by one who signed his paper, "So and So, Atheist," "What would you say if you saw a duck come out of a hen's egg?" I recognised at once the idea at the back of the question and appreciated the fact that it had been asked by one who, as some one has said, "called himself an advanced free-thinker, but was really a very ignorant and vulgar person who was suffering from a surfeit of the ideas of certain people cleverer than himself." But, as a full discussion of the matter would have taken at least as long as the lecture which I had just concluded, my reply was that before I attempted to explain it I would wait to see the duck come out of the hen's egg, since no man had as yet witnessed such an event. I do not know whether my atheistical questioner was satisfied or not, but I heard no more of him. But, after all, is it not a marvellous thing that a duck never does come out of a hen's egg? If everything happens by chance, as some would [Pg 126] have us believe, why is it that a duck does not occasionally emerge from a hen's egg? Surely this is a *miraculum*, a thing to be wondered at, yet so common that it goes unnoticed, like many other wonderful things which are also matters of common everyday occurrence, such as the spinning of the earth on its own axis and its course round the sun and through the heavens.

If we pursue this question further we shall begin to remember that creatures more nearly related to one another also "breed true." The hen and the duck are both birds, but they are not so nearly allied to one another as the lion and the tiger, both of which are *Felidæ*, or cats. Yet no one ever expects that a tiger will be born of a lioness, or *vice versa*. Further, the pug and the greyhound are both of them dogs: the name *canis domesticus* applies to both, and one would be distinguished from the other in a scientific list as "Var. (*i.e.* variety) 'pug,'" or "Var. 'greyhound.'" Yet one can imagine the surprise of a breeder if a greyhound was born in his carefully selected and guarded kennel of pugs. In a word, not only species, but varieties do tend to breed true; the child does resemble its parent or parents. No doubt the resemblance is not absolute: there is variation as well as inheritance. Sometimes the variation may be recognised as a feature possessed by a grandparent or even by some collateral relative such as an uncle or great-uncle; sometimes this may not be the case, though the non-recognition of the likeness does not in any way preclude the possibility that [Pg 127] the peculiarity may have been also possessed by some other member of the family. But on the whole the offspring does closely resemble its parents; that is to say, not only the species and the variety but the individual "breeds true." "Look like dey are bleedzed to take atter der pa," as Uncle Remus said when he was explaining how the rabbit comes to have a bob-tail. Moreover this resemblance is not merely in the great general features. Apart from monstrosities, the children of human beings are human beings; the children of white parents have white skins, those of black progenitors are black. Commonly, though not always by any means, the children of dark-haired parents are themselves dark-haired, and so on. But smaller features are also transmitted, and transmitted too for many generations; for example, the well-known case of the Hapsburg lip, visible in so many portraits of Spanish monarchs and their near relatives, and visible in life to-day. Again, there are families in which the inner part of one eyebrow has the hairs growing upwards instead of in the ordinary way, a feature which is handed on from one generation to another. Even more minute features than this have been known to be transmissible and transmitted, such as a tiny pit in the skin on the ear or on the face. In fact, there is hardly any feature, no matter how small, which may not become a hereditary possession.

If in-and-in breeding occur, as it may do amongst human beings in a locality much removed from other places of habitation, it may even [Pg 128] happen that what may be looked upon as a variety of the human race may arise, though when it arises it is always easy to wipe it out and restore things to the normal by the introduction of fresh blood, to use the misleading term commonly employed, where the Biblical word "seed" comes much nearer to the facts.

Thus there is a well-authenticated case in France (in Brittany if I remember right) of a six-fingered race which existed for a number of generations in a very isolated place and was restored to five-fingeredness when an increase in the populousness of the district permitted a wider selection in the matter of marriages.

And similarly, not long ago an account was published of an albino race somewhere in Canada which had acquired a special name.

Perhaps it has been wiped out by this time by wider marriages, though these might be effected with greater difficulty by albinos than by six-fingered persons. At any rate no one can doubt that it might at any time be wiped out by such marriages, though even when apparently wiped out, sporadic cases might be expected to occur: what the breeders call "throws-back," when they see an animal which resembles some ancestor further back in the line of descent than its actual progenitors. Certainly the most remarkable instance of the reliance which we have come to feel respecting this matter of inheritance is that which was afforded by a recent case of disputed paternity interesting on both sides of the Atlantic, since the events in dispute occurred in America and [Pg 129] the property and the dispute concerning it were in England.

It was obviously a most difficult and disputable case, but the judge, a shrewd observer, noticed, when the putative father was in the box, a feature in his countenance which seemed closely to resemble what was to be seen in the child which he claimed to be his own. A careful examination of the parents and of the child was made by an eminent sculptor, accustomed to minute observation of small features of variety in those sitting to him as models.

He reported and showed to the court that there were remarkable features in the head of the child which resembled, on the one hand an unusual configuration in the mother—or the woman who claimed to be the mother—and on the other a well-marked feature in her husband. And as a result the father and mother won their case, and were proclaimed the parents of the child because of the resemblance of these features; and, if we think for a moment, we shall see, because also of the reliance which the human race has come to place in the fidelity of inheritance, of its perfect certainty, so to speak, that a duck will not come out of a hen's egg, and the fact of this reliance on a generally received truth remains, whatever may be said as to the legal aspect of such evidence.

Inheritance is a fact recognised by everybody, and the only reason why we refuse to wonder at it is because, like other wonderful yet everyday facts, such as the growth of a great tree from a [Pg 130] tiny seed, it *is* so everyday that we have ceased to wonder at it. It is there: we know that. But have we any kind of idea how it comes about? The duck does not, as a matter of common experience, come out of a hen's egg. Why does it come out of a duck's egg? Why doesn't it come out, if only rarely, from a hen's egg? In other words, do we know what it is that explains inheritance or how it is that there is such a thing as inheritance? Well, candour obliges me to say that we do not. In spite of all the work which has been expended upon this question we are totally ignorant of the mechanism of heredity. Nevertheless it will be instructive to glance at the theories which have been put forward to explain this matter.

All living things spring from a small germ, and in the vast majority of cases this germ is the product in part of the male and in part of the female parent. It is therefore natural that we should in the first place turn our attention to this germ and ask ourselves whether there is anything in its construction which will give us the key of the mystery. There is not, at least there is nothing definite as shown by our most powerful microscopes. To be sure there is a remarkable substance, called chromatin because of its capacity for taking up certain dyes, which evidently plays some profoundly important part in the processes of development. We may suspect that this is the thing which carries the physical characteristics from one generation to another, but we cannot prove it; and though some authorities think that [Pg 131] it is, others deny it. Even if it be, it can hardly be supposed that microscopic research will ever be able to establish the fact, and that for reasons which must now be explained.

Let us suppose that we visit a vast botanic garden, and in the seed-time of each of the plants therein contained select from each plant a single ripe seed. It is clear that, if we take home that collection of seeds, we shall have in them a miniature picture of the garden from which they were culled, or at least we shall be in possession of the potentiality of such a garden, for, if we sow these seeds and have the good fortune to see them all develop, take root and grow, we shall actually possess a replica of the garden from which they came. Not exactly, it may be urged, for the distribution or arrangement of the seeds must have been carefully looked to, if the gardens are to resemble each other otherwise than in the mere possession of identical plants. I admit the truth of this, but cannot for the moment discuss it. At any rate we should have the same plants in both gardens.

On this analogy, many have suggested that every organ in the body—we must go further, and say that every marked feature in every organ in the body—is represented in the germ by a seed which can grow, under favourable circumstances, into just such another organ or feature of an organ. This was the theory put forward by Darwin under the name of "pangenesis," and by others under other titles with which it is unnecessary to burden these pages. All these theories have [Pg 132] been summed together under the name "micromeristic," that is small-fragmented, or again, "particulate," since they all postulate the existence in the germ of innumerable small fragments—seeds—which are capable of growing into complete plants or organs under favourable circumstances. Again, this, even if true, does not by any means exhaust the matter, for it does not explain why the seed of the eye implants itself and grows in the right place in the head instead of making a home for itself, let us say, in the sole of the foot. But again we must pass over that matter.

There is nothing inherently impossible in this theory; indeed, if we allow that the transmission of inheritable characteristics is purely material, and it may be, there is only one other conceivable way in which it can occur. It is true that the seeds must be almost innumerable, but the germ, though small, is capable of accommodating an almost innumerable number of independent factors, if the prevalent views as to the constitution of matter are to be believed. And, as it is quite inconceivable that we can ever have microscopes which could detect such minute objects as the ultimate bricks of which the atom—no, not even the atoms themselves which compose the germ—consists, it is impossible that we should be able to say that the seed-theory is untrue. Even if we could see these ultimate constituents it is in the last degree unlikely that they would have any resemblance to the things which are, on this theory to grow from them, [Pg 133] any more than the acorn resembles the oak which is to spring from it.

But observe! the germ on this view must contain not only seeds from the immediate parents but from many, perhaps all, of the older generations of the family, otherwise how are we to account for the appearance of ancestral peculiarities which the father and mother do not show? Moreover, since very minute things, like the inner angle of the eyebrow, may independently vary, there must be an enormous number of seeds apart altogether from the considerations alluded to in the last paragraph. And many authorities who have closely considered the question have come to the conclusion that the complexities introduced would be so great that it is impossible to believe in any micromeristic theory.

Then, of course, we must look out for some other explanation, and some have suggested that it is to be found in memory—the memory of the germ of what it was once part and the anticipation of what it may once more be. This again is an explanation not susceptible of proof along the lines of a chemical experiment, but not necessarily, therefore, untrue. Of course there are two ideas as to memory. If we are pure materialists and imagine every memory in our possession as something stamped, in some wholly incomprehensible manner, on some cell of our brain and looked at there, by some wholly inconceivable agency, when we sit down to think of past days, then we must look on the germ, under the "mnemic" or memory theory as consisting of fragments each of [Pg 134] them impressed with the "memory" of some particular organ or feature of the body, and lo! we find ourselves back again in micromerism. If we are to take a non-materialistic view of memory we are plunged into a metaphysical discussion which cannot here be pursued. A third explanation, which by the way explains nothing, is that the whole matter is one of "arrangement," to which we shall return at the close of this paper.

The mechanism of inheritance must either be physical [33] or it must be non-physical; that is, immaterial. This is what emerges from our discussion, and so far as science goes to-day it must be admitted that neither of these explanations can be said to be accepted generally by men of science or proved—perhaps even capable of proof—by scientific methods. If we know little or nothing about the mechanism of inheritance, can we and do we know anything about the laws under which it works, or has it any laws? Or are its operations a mere chance-medley? It is hardly necessary to ask the latter question, for chance-medley could not lead to regular operations— operations so regular that a court of law may act upon their evidence. Yes: we answer to the first question very lightly but without perhaps always thinking what that affirmative answer implies, a point to be considered in a moment. It may at [Pg 135] once be said that we do now know a good deal about the laws under which inheritance works itself out, and that knowledge, as most people are now aware, is due to the quiet and for a time forgotten labours of Johann Gregor Mendel, once Abbot of the Augustinian Abbey of Brünn, a prelate of that Church which loud-voiced ignoramuses are never tired of proclaiming to have been from the beginning even down to the present day the impassioned and deadly enemy of all scientific progress. Mendel saw that former workers at inheritance had been directing their attention to the *tout ensemble* of an individual or natural object; his idea was analytical in its nature, for he directed his attention to individual characteristics, such as stature or colour, or the like. And having thus directed his attention and confined his labours mainly to plants, since the study of generations of most animals is too lengthy a process for one man to carry out, he did in fact discover that there are very definite laws, capable even of numerical statement, under which inheritance acts. There is no need to explain or discuss them here: suffice it to say that there *are* such laws, [34] as is now admitted by an overwhelming majority of the biologists of to-day. Mendel's facts were hidden in a somewhat obscure journal; they lay dormant, much to his annoyance, during his lifetime. Years after his death his papers were unearthed, and his [Pg 136] discoveries have been proclaimed as being as fundamental to biology as those of Newton and Dalton to other sciences.

There are, then, laws. That means one of two things: either that these laws arose by chance-medley, or that some one enacted them. It seems impossible, when one surveys the orderly operations of Nature, among which are those conducted under the laws known by the name of their discoverer, Mendel — it seems wholly impossible that these operations arose by chance-medley. To me, at any rate, any such explanation is wholly unthinkable. But if it be an impossible explanation, as I and many thousands, not to say millions, of other persons believe, then there is no other way out of it than that these operations must have been planned by some one; in other words, that there must have been a Creator and Deviser of the world.

People hide from this explanation, and one of the favourite sandbanks in which this particular kind of human ostrich plunges its head is "Nature." "Nature does this," and "Nature does that," forgetting entirely the fact that "Nature" is a mere personification and means either chance-medley or a Creator, according to the old dilemma. There is a very curious example of this inability or unwillingness to admit—perhaps even to understand—the force of this argument exhibited by those to whom one would suppose that it would come home with overpowering force: I mean, of course, the Mendelians.

The most learned of these, and one of the most [Pg 137] open-minded of men, hints in one place that though he does not think it necessary himself to believe it, yet it might at least be suggested that, if in a certain organism we find things so placed that a certain combination is bound to emerge in a certain generation, such a state of affairs might have been prearranged. Now, if it was prearranged, the awful fact emerges that there must have been an arranger; in other words, a creative power. This explanation is taboo in certain circles. But one may reasonably ask, "What then?" Is it really suggested that these orderly sets of occurrences may occur not once or twice only but thousands and thousands of times, and this may all happen by chance? A very distant acquaintance with the mathematics of probability will show that this is a wholly untenable theory. We are generally answered by some purely verbal explanation, like the personification of "Nature" already alluded to.

Thus, in a recent discussion on inheritance in a Presidential Address to the British Association, to which I have already alluded, the writer with whose explanation I have just been dealing states that he thinks it "unlikely" that the factors of inheritance are "in any simple or literal sense material particles," and proceeds thus: "I suspect rather that their properties depend on some phenomenon of arrangement." Now, in the first place, this is no explanation at all, for the mechanism of inheritance must be either material or immaterial. If there is a phenomenon of "arrangement" there must be something to be [Pg 138] "arranged," and this something can hardly be other than material if it is to be "arranged" at all. But let that pass. What is far more important is to remember that if a thing is to be "arranged" there must be somebody to "arrange" it, for chance-medley cannot "arrange" anything in an orderly manner; or if it could do so once, cannot be supposed capable of doing it a second time in a precisely similar manner, not to say capable of doing it countless thousands of times.

If we go into a great museum our first idea, perhaps our last, concerns the arrangement found therein. But it may safely be said that no sane person ever entertained that idea without being perfectly aware that the arrangement was made by human hands, controlled, in the last resort, by the brain of the curator of the museum. Now, in a sense, the living body is a museum containing specimens of different kinds of cells. There are brain-cells, liver-cells, bone-cells, scores of different varieties of cells, and all of them, so to speak, are arranged in their appropriate cases.

If we go to the brain-case we can search it through and through without finding a liver-cell, any more than we should find a typical brain-cell embedded in the marrow of one of the bones. The different specimens all occupy their appropriate positions. How did they get there? The future animal, like animals of all kinds, including man, commences as a single cell. All save a few interesting but at present negligible cases are composed of elements drawn from male [Pg 139] and female parents. This cell divides up into a multitude of others. At first these are to all appearances identical, but later they begin to differentiate, at first into three classes and afterwards into the multitude of different cells of which the body is composed. Further, these groups of cells become aggregated in appropriate groups, cells of one kind uniting with cells of the same kind and with no others. Here we have to do with arrangement, consummately skilful arrangement, an arrangement which practically never fails, for, leaving aside the case of monstrosity, a consideration of which would detain us too long, not merely are the various cells all placed in their proper positions, as we have seen, but their aggregation, the individual, is so formed as to belong to the proper compartment of that large museum, the world—the same compartment as that occupied by his progenitors. Neither the particulate nor the chemical theories help us here. The mnemic would, but it has its initial and insuperable difficulty, pointed out in another article in this volume, that, as you must have an experience before you can remember it, it in no way accounts for the first operation of arrangement. As to the material explanations, particulate or chemical, they amount to something like this: you have half a cart-load of bricks from one yard and half a cart-load from another, and when the bricks are dumped down in an appropriate place they form a little house, just like those occupied by the managers of the brickyards. So they may, but no one in his sense supposes that [Pg 140] they will thus arrange themselves of their own power. Some one must arrange them. Who arranges the tiny bricks of which the animal body consists, or what arranges them? To revert to our previous example of the garden; suppose that we bring back from that which we desire to copy a bag of seeds representing all the plants which it contains. We have a plot of land of the same size as our example; we dig it and we dung it and then we scatter our seeds perfectly haphazard over its surface. What are the odds as to their coming up in an exactly similar pattern to those in the other garden. Mathematicians, I suppose, could calculate the probabilities, but they must be infinitesimally small. Yet in the case of the animal the pattern is always observed.

It is quite useless for any one, however eminent an authority he may be, to dismiss the matter by saying "It is a phenomenon of arrangement," for that begs the whole question. A Martian visitor taken to Westminster Abbey and told that its construction was a "phenomenon of arrangement" might be expected to turn a scornful eye upon his cicerone and reply, "Any fool can see that, but who arranged it?"

Hence, though wild horses would not drag such an admission from many, we are irresistibly compelled to adopt the theory of a Creator and a Maintainer also of nature and its operations—so-called—if we are to escape from the absurdities involved in any other explanation. Thus there are very important and fundamental matters to be deduced from the very little which we know [Pg 141] about inheritance, just as there are from a hundred and one other lines of consideration related to this world and its contents. We do not know very much—it may fairly be said we *know* nothing as to the vehicle of inheritance. We know a little, but it is still a very little even in comparison with what we may yet come to know as the result of careful and long-continued experiment, about the laws of inheritance. What we do learn from our knowledge, such as it is, is the fact that we can give no intelligent or intelligible explanation of the facts brought before us except on the hypothesis of a Creator and Maintainer of all things.

FOOTNOTES:

[Pg 142]

[33] A third explanation, that the mechanism of inheritance is of a chemical character, is now being put forward, and some mention of this view, which is by no means one of general acceptance, will be found in another article in this volume.

[34] An account of them will be found in *A Century of Scientific Thought*, by the present writer, published by Messrs. Burns & Oates.

VII. "SPECIAL CREATION"

Professor Scott, of Princeton, has recently given to the public in his Westbrook Lectures [35] an exceedingly impartial, convincing, and lucid statement of the evidence for the theory of evolution or transformism. On one point of terminology a few observations may not be amiss, since there is a certain amount of confusion still existing in the minds of many persons which can be and ought to be cleared up. Throughout his book Professor Scott contrasts evolution with what he calls "special creation." In so doing he is evidently in no way anxious to deny the fact that there is a Creator, and that evolution may fairly be regarded as His method of creation. In one passage he expressly states that "acceptance of the theory of evolution by no means excludes belief in a creative plan."

And again, when dealing with the palæontological evidence in favour of evolution, he points out that Cuvier and Agassiz, examining it as it was known in their day, interpreted the facts as [Pg 143] the carrying out of a systematic creative plan, an interpretation which the author claims "is not at all invalidated by the acceptance of the evolutionary theory." He is not, we need hardly say, in any way singular in taking up this attitude, since it was held by Darwin, by Wallace, by Huxley, and by other sturdy defenders of the doctrine of evolution.

Yet, just as at the time that Darwin's views were first made public, many thought that they were subversive of Christianity, so, even now, some whose acquaintance with the problem and its history is of a superficial character, are inclined when they see the word creation, even with the qualifying adjective "special" prefixed to it, used in contradistinction to evolution, to imagine that the theory of creation, and of course of a Creator, must fall to the ground if evolution should be proved to be the true explanation of living things and their diversities.

It is more than a little difficult for us, living at the present day, to understand this curious frame of mind; yet it certainly existed, and existed where it might least have been expected to exist. Nor is it quite extinct to-day, though it only lingers in the less instructed class of persons. The misconception arose from a confusion between the fact and the method of creation. As to the former, no Catholic, no Christian, no theist has any kind of doubt; indeed there are those who could not be classified under any of those categories who still would be prepared to admit that there must be a First Cause as the [Pg 144] explanation of the universe. Some of them, whose reasoning is a little difficult to follow, seem to be content with an immanent, blind god, a mere mainspring to the clock, making it move, no doubt, but otherwise powerless. If we neglect—in a mathematical sense—those who adopt the agnostic attitude; content themselves with the formula *ignoramus et ignorabimus* of Du Bois Reymond, and confine their investigations to the machine as a going machine without inquiring how it came to be a machine or what set it to work, we shall, I think, find that most people who have really thought out the question admit that the only reasonable explanation of things as they are, is the postulation of a Free First Cause; in other words, an Omnipotent Creator of the universe. Such, of course, is the teaching of the Scriptures and of the Church, and it must be admitted that neither of them carries us very much further in this matter. In fact, whilst both are perfectly clear and definite about the fact of creation, neither of them has much to say about the method. Yet, as all admit, evolution concerns only the method and tells us absolutely nothing about the cause.

Being omnipotent, it is obvious that its Maker might have created the universe in any way which seemed good to Him—for example, all at once out of nothing just as it stands at this moment. Such a thing would not be impossible to Omnipotence; and, as we know,

Fallopius, suddenly confronted by the problems of fossils in the sixteenth century, did suggest that they were [Pg 145] created just as they were, and that they had never been anything else. So did Philip Gosse some two and a half centuries later.

There is nothing more sure than that the world was not created just as it is. Reason and Scripture both teach us that, and geology makes it quite clear that the appearance of living things upon the earth has been successive; that groups of living things, like the giant saurians, which were once the dominant zoological objects, had their day and have gone, as we may suppose, for ever. A few very lowly forms, like the lamp-shells, have persisted almost throughout the history of life on the earth, but on the whole the picture which we see is one of appearances, culminations, and disappearances of successive races of living things. There was a time when Trilobites, crustaceans whose nearest living representatives are the King-Crabs, first became features of the fauna of the earth. Then they increased to such an extent as to become the most prominent feature. Then they declined in importance, disappeared, and for uncounted ages have existed only as fossils. Thus we conclude that the creation of species was a progressive affair, just as the creation of individuals is a successive affair, for every living thing, coming as it does into existence by the power of the Creator, is His creation and in a very real sense a special creation. Now we know very well how living things come into existence to-day; can we form any idea as to how they originated in the beginning? Milton, in his crude description in *Paradise Lost*, pictured living things as gradually [Pg 146] rising out of and extricating themselves from the soil.

> "The grassy clods now calved, now half appeared
> The tawny lion, pawing to get free
> His hinder parts, then springs as broke from bonds,
> And rampant shakes his brindled mane; the ounce,
> The libbard, and the tiger, as the mole
> Rising, the crumbled earth above them threw
> In hillocks: the swift stag from underground
> Bore up his branching head: scarce from his mould
> Behemoth, biggest born of earth, up heaved
> His vastness."

In this description Milton probably represented the ideas of his day—a day penetrated with literal interpretation of the Scripture, though it is well to recall to our minds the fact that not one word or idea of the above is contained in the Bible. The only suggestion is that the body of Adam was fashioned from the "slime of the earth," the precise meaning of which phrase has never been defined by the Church.

Again, we have to say that the Miltonic scheme is not impossible, any more than any other scheme is impossible, but we may further say that it is more than improbable, and with every reverence we may add that to us it does not seem to be specially consonant with the greatness and wisdom of God. There remains the derivative form of creation, compendiously styled evolution. That this also is a possible method of creation no one will deny, and it has been discussed as such by many of the greatest thinkers in the history of the Church. We can consider it, [Pg 147] therefore, from the point of fact or of knowledge as we now possess it, and we can do so without imagining that, in so doing, we are contemplating a method which is anything else but the carrying out of a creative plan, existing perfect and complete and from all eternity in the mind of the Being Whose conception it was and by whose *fiat* it came to pass. Moreover, each form produced is a special creation, since it was specially designed to be as it is and to appear when it did, just as the clockmaker intends his clock to strike twelve at noon, though he can hardly be said to make it strike at that moment. Hence to place special creation in antagonism to evolution is really to use an ambiguous phraseology. No doubt it is not easy to find the proper phraseology. Some have employed the terms "immediate" and "mediate," to which also a certain amount of ambiguity is attached. Perhaps "direct" and "derivative" might convey more accurate ideas; but whatever terminology we adopt, we are still safe in saying that whether God makes things or makes them make themselves He is creating them and specially creating them.

This is not the place to enter into any elaborate discussion as to the truth of the theory of evolution. Few will be found to deny the statement that it is a theory which *does* explain Nature as we see it and as we learn its history in the past, but that does not necessarily prove that it is true. St. Thomas Aquinas, dealing with the movements of the planets, makes a very important statement when he tells us, in so many words, that, though [Pg 148] the hypothesis with which he is dealing would explain the appearances which he was seeking to explain, that does not prove that it is the true explanation, since the real answer to the riddle may be one then unknown to him. There are, however, one or two points it may be useful to consider before we leave the question.

That evolution may occur within a class seems to be quite certain. The case of the Porto Santo rabbits, one of many cited by Darwin or brought to knowledge since his time, will make clear what is meant. Porto Santo is a small island, not far from Madeira, on which a Portuguese navigator, named Zarco, let loose, somewhere about the year 1420, a doe and a recently born litter of rabbits, which we may feel quite sure belonged to one of those domestic breeds which have all been derived from the wild rabbit of Europe known to zoologists as *Lepus Cuniculus*. The island was a favourable spot for the rabbits, for there do not appear to have been any carnivorous beasts or birds to harry them, nor were there other land mammals competing with them for food; and, as a result, we are told that they had so far increased and multiplied in forty years as to be described as "innumerable." In four and a half centuries these rabbits had become so different from any European rabbits that Haeckel described them as a species apart, and named it *Lepus Huxlei*. This rabbit is much smaller than the European form, being described as more like a large rat than a rabbit. Its colour is very different from its European [Pg 149] relatives; it has curious nocturnal habits; it is exceedingly wild and untamable. Most remarkable of all, and most conclusive as to specific difference, Mr. Bartlett, the highly skilled head keeper of the London Zoological Gardens, utterly failed to induce the two males which were brought over to those gardens to associate with or to breed with the females of various other breeds of rabbits which were repeatedly placed with them. If the history of these Porto Santo rabbits had been unknown to us, instead of being a matter as to which there can be no doubt, every naturalist would at once have accepted them as a separate species. We need not hesitate, it appears, to do so and to admit that it is a new species which has been produced within historic times and under conditions with which we are fully acquainted. It may, however, be argued, and quite fairly argued, that such a process of evolution, though definitely proved, is a very different thing from such an evolution as would permit of a common ancestry for animals so far apart, for example, as a whale and a rabbit, or perhaps even nearer in relationship, as between a lion and a seal. To discuss this further would require a dissertation on the highly involved question of species and varieties, and that is not now to be attempted. What, however, may be said is that the difficulties presented by what is called phylogeny — that is, the relationships of different classes to one another — are so great as to have led more than one man of science to proclaim his belief that evolution has been poly— [Pg 150] and not mono— phyletic. Such is the view which has been enunciated by Father Wasmann, S.J., whose authority on a point of this kind is paramount. It has also been upheld by Professor Bateson, a man widely separated from the Jesuit in all but attachment to science. Professor Bateson summed up his belief in the text which he placed on the title-page of his first great work on *Variation*: the text which proclaims that there is a flesh of men, another of beasts, another of birds, another of fishes.

Darwin remained to the end of his life undecided between the two views, for he allowed his original statement as to life having been breathed into one or more forms by the Creator, to pass from edition to edition of the *Origin of Species*. If the polyphyletic theory be adopted, it must be said that the position of the materialist is made far more difficult than it is at present. Let us see what it means. On the materialistic hypothesis, and the same may be said of the pantheistic or any other hypothesis not theistic in nature, a certain cell came by chance to acquire the attributes of life. From this descended plants and animals of all kinds in divergent series till the edifice was crowned by man. I have elsewhere endeavoured to point out all that is involved in this assumption, which, it must be confessed, is a very large mouthful to swallow.

Let us now consider what the polyphyletic hypothesis involves. According to this view one cell accidentally developed the attributes of vegetable life; a further accident leads another [Pg 151] cell to initiate the line of invertebrates; another that of fishes, let us say; another of mammals: the number varying according to the views of the theorist on phylogeny. Let us not forget that the cell or cells which accidentally acquired the attributes of life, had accidentally to shape themselves from dead materials into something of a character wholly unknown in the inorganic world. If one seriously considers the matter it is—so it seems to me—utterly impossible to subscribe to the accidental theory of which the immanent god—the blind god of Bergson—is a mere variant. One must agree with the late Lord Kelvin that "science positively affirms creative power ... which (she) compels us to accept as an article of belief." But what are we to say with regard to the series of repeated accidents which the polyphyletic hypothesis would seem to demand? Is it really possible that any man could bring himself to place credence in such a marvellous series of occurrences? Monophyletic or polyphyletic evolution, whichever, if either, it may have been, presents no difficulty on the creation hypothesis.

The Divine plan might have embraced either method. It is not merely revelation but ordinary reason which shows us that the wonderful things which we know, not to speak of the far more wonderful things at which we can only guess, cannot possibly be explained on any other hypothesis than that of a Free First Cause—a Creator.

FOOTNOTES:

[Pg 152]

[35] *The Theory of Evolution.* By William Berryman Scott. New York: The Macmillan Co.

VIII. CATHOLIC WRITERS AND SPONTANEOUS GENERATION

The names of great Catholic men of science, laymen like Pasteur and Müller, or ecclesiastics like Stensen and Mendel, are familiar to all educated persons. But even educated persons, or at least a great majority of them, are quite ignorant of the goodly band of workers in science who were devout children of the Church. Nothing perhaps more fully exemplifies this than the history of the controversy respecting the subject whose name is set down as the title of this paper. For centuries a controversy raged at intervals around the question of spontaneous generation. Did living things originate, not merely in the past but every day, from non-living matter? When we consider such things as the once mysterious appearance of maggots in meat it is not wonderful that in the days before the microscope the answer was in the affirmative.

To-day the question may be considered almost closed. True, the negative proposition cannot be proved, hence it is impossible to say that spontaneous generation does not take place. However, the scientific world is at one in the [Pg 153] belief that so far all attempts to prove it have failed utterly.

St. Thomas Aquinas had a celebrated and sometimes misunderstood controversy with Avicenna, a very famous Arabian philosopher. It was a philosophical, but not strictly scientific, controversy, for both persons accepted or assumed the existence of spontaneous generation. Avicenna claimed that it took place by the powers of Nature alone, whilst St. Thomas adopted the attitude which we should adopt to-day, were spontaneous generation shown to be a fact, namely, that if Nature possessed this power, it was because the Creator had willed it so.

We come to close quarters with the question itself in 1668, when Francesco Redi (1626-1697) published his book on the generation of insects and showed that meat protected from flies by wire gauze or parchment did not develop maggots, whilst meat left unprotected did. From this and from other experiments he was led to formulate the theory that in all cases of apparent production of life from dead matter the real explanation was that living germs from outside had been introduced into it. For a long time this view held the field. Redi was, as his name indicates, an Italian, an inhabitant of Aretino, a poet as well as a physician and scientific worker. He was physician to two of the Grand Dukes of Tuscany and an academician of the celebrated *Accademia della Crusca*. Those works which I have been able to consult on the subject say nothing about his religion, but there can scarcely [Pg 154] be any doubt that he was a Catholic. At any rate there is no doubt whatever as to the other persons now to be mentioned in connection with the controversy, which again became active about a century after Redi had published his book. The antagonists on this occasion were both of them Catholic priests, and both of them deserve some brief notice.

John Turberville Needham (1713-1781) was born in London and belonged on both sides to old Catholic families. He was educated at Douay and ordained priest at Cambray in 1738. After teaching in that place for some time he journeyed to England and became headmaster of the once celebrated school for Catholic boys at Twyford, near Winchester. From there he went for a short time to Lisbon as professor of philosophy in the English College. Subsequently he travelled with various Peers making "the grand tour." After that he retired to Paris, where he was elected a member of the *Académie des Sciences*. He was the first director of the Imperial Academy in Brussels; a canon, first of Dendermonde and afterward of Soignies. He died in Brussels and was buried in the Abbey of Condenberg. Needham was a man of really great scientific attainments, and perhaps nothing proves the estimation in which he was held more than the fact that in 1746 he was elected a Fellow of the Royal Society, being the first Catholic priest to become a member of that distinguished body. When one remembers the attitude at that time, and much later, of Englishmen towards [Pg 155] Catholics it is clear that Needham's claims to distinction must have been more than ordinarily great. His clear, firm signature is still to be seen in the charter-book of the society, and it is interesting to note that he signs his name "Turberville Needham." Needham did not confine his attention to science, for he was an ardent antiquary, and in 1761 was elected a Fellow of that other ancient and exclusive body, the Society of Antiquaries of London. In this connection it may be mentioned that Needham published, in 1761, a book which caused a great sensation, for he endeavoured to show that he could translate an Egyptian inscription by means of Chinese characters; in other words, that the forms of writing were germane to one another. He was shown to be quite wrong by some of the learned Jesuits of the day, who, with the assistance of Chinese men of letters, proved that the resemblances to which Needham had called attention were merely superficial.

But our interest now is in his controversy with Spallanzani. Lazaro Spallanzani (1729-1799) was born at Scandiano in Modena and educated at the Jesuit College at Reggio di Modena. There was some question as to his entering the Society; he did not do so, however, but repaired to the University of Bologna, where his kinswoman,

Laura Bassi, was then professor of physics. He became a priest, but devoted his life to teaching and experimenting. He must have been something of what we in Ireland used to call a "polymath," for he professed at one time or another, [Pg 156] in various universities, logic, metaphysics, Greek, and finally natural history. He first explained the physics of what children call "ducks and drakes" made by flat pebbles on water; laid the foundations of meteorology and vulcanology, and is perhaps best of all known in connection with what is termed "regeneration" in the earthworm and above all in the salamander. His experiments still hold the field in a region of study which has vastly extended itself in recent years, becoming of prime importance in the vitalistic controversy.

In the dispute, however, with which we are concerned Needham and Spallanzani defended opposite positions. The former, as the result of his observations, asserted that, in spite of the boiling and sealing up of organic fluids, life did appear in them. His opponent claimed that Needham's experiments had not been sufficiently precise. The latter had enclosed his fluids in bottles fitted with ordinary corks, covered with mastic varnish, whilst Spallanzani, employing flasks with long necks which he could and did seal by heat when the contents were boiling, showed that in that case no life was produced. He declared, and correctly too, as we now know, that Needham's methods did permit of the introduction of something from without. The controversy went to sleep again until the discovery of oxygen by Priestley in 1774. When it had been shown that oxygen was essential to the existence of all forms of life, the question arose as to whether the boiling of the organic fluids in the earlier [Pg 157] experiments had not expelled all the oxygen and thus prevented the existence and development of any life.

In the further experiments which this query gave rise to, we meet with another illustrious Catholic name, that of Theodor Schwann, better known as the originator of that fundamental piece of scientific knowledge, the cell-theory. Theodor Schwann (1810-1882) was born at Neuss and educated by the Jesuits, first at Cologne, afterward at Bonn. After studying at the Universities of Würzburg and Berlin he became professor in the Catholic University of Louvain, where his name was one of the principal glories of this now wrecked seat of learning. Thence he went as professor to Liége, where he died. He was, says his biography in the *Encyclopædia Britannica*, "of a peculiarly gentle and amiable character and remained a devout Catholic throughout his life." Schwann's experiments tended to show that the introduction of air—of course containing oxygen—did not lead to the production of life, if the air had first been thoroughly sterilised. It was thought that this question had been finally answered, when it was reopened by Pouchet, in 1859. He was a Frenchman, the director of the Natural History Museum of Rouen, but as to his religious views I have no information. It is quite probable, however, that he was a Catholic. Pouchet and all on his side were finally—so far as there can be finality in such a matter—disposed of by Pasteur, of whose distinction as a man of [Pg 158] science and devoutness as a Catholic nothing need be said.

It is quite unnecessary to devote any consideration here to the character of Pasteur's experiments, for they have become a matter of common knowledge to all educated persons. Let it suffice to say that they were on the lines first laid down by Redi and greatly elaborated by Spallanzani, namely the exclusion from the fluids or other substances under examination of all possible contamination by minute organisms in the air. Spallanzani knew nothing of these organisms; they were not discovered until many years after his death. But he surmised that there was something which brought corruption into the fluids; he excluded that something, with the result that the fluids remained untainted. From our point of view, however, there are several things to be learnt. In the first place quite a number of ignorant persons have thought that the discovery of spontaneous generation would upset religious dogmata. That of course is quite absurd. From what has been said above it will be seen that St. Thomas Aquinas—in common with all the men of learning of his day—fully believed in it, as did Needham, another ecclesiastic as to whose orthodoxy there is no doubt. Further, the entire controversy is a complete confutation of the false allegation that between Catholicism and science there is a great gulf set. There have been few longer and more remarkable controversies in the history of science, and scarce any other—if indeed any other—which has such important bearings [Pg 159] upon health and industry than that which relates to bio- or abio-genesis. It is significant to find that the names of so many of the protagonists in this controversy were those of men who were also convinced adherents of the Catholic Church.

[Pg 160]

IX. A THEORY OF LIFE [36]

Of the making of books on the question of Vitalism there would seem to be no end; and, following upon quite a number of others comes this handsome, well-illustrated, intensely interesting book, by one whose writings are always worth study. It purports to deal with the Origin and Evolution of Life; but, as to the first, it leaves us in no way advanced towards any real explanation of that problem on materialistic lines. As to the second, though there is a vast amount of valuable information, often illuminating and suggestive, again we confess that we fail to discover any real philosophy of that process of evolution which the author postulates. These propositions we must now proceed to justify. We can consider them from the most rigidly scientific standpoint, since, if every word or almost every word in the book were proved truth, it would not make the slightest difference to Catholic Philosophy, nor, indeed, to Theistic teachings, since in the imperishable words of Paley: "There may be many second causes, and many courses of second causes, one behind another, [Pg 161] between what we observe of nature and the Deity; but there must be intelligence somewhere; there must be more in nature than what we see; and, amongst the things unseen, there must be an intelligent designing Author."

The scientific writer has to remember that whilst he may explain many things, his work is a torso unless and until he has either accepted the Creator as the first Cause, which he is too often disinclined to do, or has supplied an equally satisfactory explanation, which he is permanently unable to do. On the other hand, at least some defenders of Theism in the past might well have borne in mind that, whilst we are assured of the fact of Creation, we know absolutely nothing of its mechanism save that it came about by the command of God. There is nothing in which clear thinking and clear writing are more necessary than in discussions of this kind; and too many of them are vitiated by an obvious lack of philosophical training on the part of the participants. Even in this carefully written book there are instances of this kind of thing to which we must allude before considering its main arguments.

"We know, for example, that there has existed a more or less complete chain of beings from monad to man, that the one-toed horse had a four-toed ancestor, that man has descended from an unknown ape-like form somewhere in the Tertiary." "We *know*"— that is exactly the opposite of the truth. We *know* a thing when it is susceptible of proof according to the rigid rules of formal logic; when, to doubt it, would be to [Pg 162] give rise to a suspicion as to our sanity; then we *know* a thing, but not until then. Now, as to the sentence quoted, we may allow the first part to pass unchallenged with some possible demur at the use of the word "chain." The second so-called piece of knowledge was doubted by no less an authority than the late Adam Sedgwick. The third assertion plainly and distinctly is not the case; for Science *knows* nothing whatsoever about the origin of man's body. In 1901 Branco, a distinguished palæontologist, with no Theistic leanings as far as we know, told the world that man appears on our planet as "a genuine *homo novus*," and that palæontology "knows no ancestors of man." Nor has any discovery since that date necessitated the modification of that opinion. What the writer means by saying "*We* know" is "*I* am convinced"; but, with the deepest respect for his undoubted position, the two things are not quite identical. "Biology, like theology, has its dogmas. Leaders have their disciples and blind followers." Wise words! They are those of the author with whom we are dealing. To say "we know" when really we only surmise is a misuse of language, just as it is also a misuse to ask the question "Does nature make a departure from its previously ordered procedure and substitute chance for law?" since the ordinary reader is all too apt to forget that "Nature" is a mere abstraction, and that to speak of Nature doing such or such a thing helps us in no way along the road towards an explanation of things.

[Pg 163]

Or again: "So far as the *creative* power of energy is concerned, we are on sure ground." The author has a careful note on the word creation (p. 5), "the production of something new out of nothing," under which definition it is abundantly clear that energy, whilst it may be *productive*, cannot be *creative*. In fact, nothing can be *creative* in any definite and rigid sense, save a *Creator* Who existed from all eternity and from Whom all things arose. One more instance of loose argumentation, and we can turn to the main purport of the book. It is a link in the author's "chain" which cannot be passed without examination. Everybody is familiar with the method of proof by elimination. We set down every possible explanation of a certain occurrence; we rule out one after the other until but one is left. If we really have set down all the possible explanations, and if we are quite clear as to the fact that all those which have been excluded are legitimately put out of court, then the one remaining explanation must be the true one. It is a method of proof which has frequently been applied to the vitalistic problem, and with the greatest effect, as it is admitted by some of those who would greatly like to find a materialistic explanation for that problem (cf. *The Philosophy of Biology*, Johnstone, p. 319).

Let us see how our author employs it. What, he asks, is "the internal moving principle" in living substance? And he replies: "We may first exclude the possibility that it acts either through supernatural or teleological interposition [Pg 164] through an externally creative power." Very well! Philosophers tell us that we can assume any position we choose for the purposes of our argument, but that ultimately we must prove that assumption or admit ourselves beaten. We look anxiously for the proof of the assumption made by our author, but absolutely no attempt is made to give one. We must be pardoned, therefore, if we hesitate to accept such an important statement on his mere *ipse dixit*. We pass on to the next elimination: "Although its visible results are in a high degree purposeful, we may also exclude as unscientific the vitalistic theory of an *entelechy* [37] or any other form of internal perfecting agency distinct from known or unknown physio-chemical energies." Why "unscientific"? Numbers of high authorities have not thought it so; and in quite recent years such eminent writers as Driesch and McDougal have written erudite works to prove this "unscientific" hypothesis. Is there any proof brought forward for *this* assertion and its corresponding elimination?

Let us continue the quotation: "Since certain forms of adaptation which were formerly mysterious can now be explained without the assumption of an entelechy we are encouraged to hope that all forms may be thus explained." The author does not tell us what the mysterious [Pg 165] adaptations are, nor does he offer us the explanations which, in his opinion, explain them. We cannot, therefore, criticise his views, and can only remind his readers that, because an explanation plausibly explains an occurrence, it is by no means always therefore certain to be the true explanation; it may, indeed, be wholly false.

Further, those who have been wandering for the past half-century in the fields of science have become a little wearied of "explanations," vaunted, for periods of five or ten years, as the key to open all locks, and then cast into the furnace. What the author would seem to mean by his statement is this: "I am convinced myself that we can do without a 'supernatural' explanation, and I regard as 'unscientific' any explanation which cannot be put to the test of chemistry and physics; hence I must shut the door on anything like an *entelechy*, and, that being so, it behoves me to look for some other explanation." Of course, we are putting these words into the mouth of our author; if we were dealing with the matter ourselves we should be inclined to argue that, by the eliminatory method, chemistry and physics do prove, or do help to prove, the existence of an entelechy.

With these expostulations we may turn to the writer's pronouncements on the vitalistic question which seem to us to be worthy of serious consideration. Everybody knows that there are two very diverse opinions on this topic; the one that there is, the other that there is not something more—a *plus*—in living than there is in not- [Pg 166] living objects. In other words, that there is a difference of kind, and not merely of degree, between a stone and a sparrow. Hence the schools of thought called vitalistic and mechanistic. To most persons it has up to now seemed impossible that there could be a third school; we appeared to be confronted with what the logicians call a Dichotomy. Professor Osborn seems to us to think otherwise, though he is not wholly clear on this matter. If we are to "reject the vitalistic hypotheses of the ancient Greeks, and the modern vitalism of Driesch, of Bergson, and of others," and if, on the other hand, we are to view, as he thinks we must, the cosmos as one of "limitless and *ordered* energy"—we have emphasised the word "*ordered*" for reasons which will shortly appear—we must clearly look out for some middle way. "*Ordered*," a purely mechanistic and materialistically realised cosmos cannot be. "*Ordered*" conditions are determined by what we agree to call "Laws"; and these, as all must admit, entail a Lawgiver.

The alternative is Blind Chance; and the author, after considering the question, agrees, as again most reasonable persons will agree, that Blind Chance is no explanation of things as they are. He quotes a modern chemist who, discussing the probability of the environmental fitness of the earth for life being a mere chance process, remarks: "There is, in truth, not one chance in countless millions of millions that the many unique properties of carbon, hydrogen, and oxygen, and especially of their stable compounds, [Pg 167] water and carbonic acid, which chiefly make up the atmosphere of a new planet, should simultaneously occur in the three elements otherwise than through the operation of a natural law which somehow connects them together. There is no greater probability that these unique properties should be without due cause uniquely favourable to the organic mechanism" (J. J. Henderson, 1913).

If neither of the classic points of view is tenable, what then is the explanation, if, indeed, any be possible? The author casts one brief glance down that blind-alley marked "Element Way." Does some known element or some unknown element, to which the name *Bion* might be given, exist and form the source of the energy in living things? Radium has only been known to us for a few years; can we say that there is no such thing as Bion? Of course we cannot; but this we can say, that, if there is such an element and if it is really responsible for all the protean manifestations of life, wonderful as radium and its doings are, they must sink into nothingness beside those of this new and unsuspected entity. The author evidently does not think that this path is a profitable one to pursue, and we agree with him; so he turns his attention to the question of energy. Energy is the capacity for doing work. It is often, of course, latent, as, for example, in a cordite cartridge, which is a peaceful, harmless thing until the energy stored up in it is realised with the accompanying explosion and work is done. It is the same with a bent spring; [Pg 168] a clock-weight when the clock is not going, and so on.

We need not develop this matter further; but one point must be alluded to, namely, the gradual exhaustion of the available energy in the changes from one manifestation to another. In all physical processes heat is evolved, which heat is distributed by conduction and radiation and tends to become universally diffused throughout space. When complete uniformity has been attained, all physical phenomena will come to an end; in other words, our solar system must come to an end, and it must have had a beginning. It is a well-known argument. Is there anything to rewind the clock which is running down before our very eyes? It was once urged that stellar collisions, and such-like things, might permit us to postulate a cyclical arrangement (and thus rearrangement) of universal phenomena; but that hypothesis does not seem to find any supporters to-day.

In his interesting book, already mentioned, Dr. Johnstone called attention to the power possessed by living matter of reversing the process; but no reversal of this kind and extent can make up for the constant degradation of energy which is taking place all round us. We mention this because it shows that "energy" cannot, in any case, afford an eternal solution, but only a temporal and therefore a limited one. No one doubts that there is energy in the living thing, nor that there are what the author calls "complexes of energies." No one, again, will [Pg 169] quarrel with the statement that energy is first seen in the sun, in the earth, in the air, and in the water; that "with life something new appears in the universe, namely, a union of the internal and external adjustment of energy which we appropriately call an *Organism*." That "the germ is an energy complex" is no doubt an unproved hypothesis, as he admits, but is quite likely. With all these assertions we may agree, though we cannot with that which follows, namely, that energy is creative, for that such is impossible in any true sense of that word we have already tried to show.

We have now to ask ourselves in what way this energy conception of life differs from, or goes beyond, the two theories of life — mechanistic and vitalistic, which have hitherto been supposed to have exhausted the possibilities of explanation. In order to do this we must analyse the author's idea of energy and its relationship to biological processes a little more closely. He begins his study of life and its evolution by considering how nutrition and the derivation of energy can have taken place before chlorophyl had come into existence; and he very pertinently points to the *prototrophic* bacteria as probably representing "the survival of a primordial stage of life chemistry." Thus a "primitive feeder," the bacterium *Nitrosomonas*, "for combustion ... takes in oxygen directly through the intermediate action of iron, phosphorus or manganese, each of the single cells being a powerful little chemical laboratory which contains oxidising catalysers, [Pg 170] the activity of which is accelerated by the presence of iron and manganese. Still, in the primordial stage, *Nitrosomonas* lives on ammonium sulphate, taking its energy (food) from the nitrogen of ammonium and forming nitrates. Living symbiotically with it is *Nitrobacter*, which takes its energy (food) from the nitrates formed by *Nitrosomonas*, oxidising them into nitrates. Thus these two species illustrate in its simplest form our law of the *interaction of an organism* (*Nitrobacter*) *with its life environment* (*Nitrosomonas*)" (p. 82, author's italics).

Once one has got to this stage, it is *ex hypothesi* easy to ascend through the vegetable and animal worlds and to formulate the various laws which appear to have shaped the evolution of life and of species. We are then "within the system," but to arrive at anything worthy of the name of an explanation we have first to *get* within the system. Even then there remains over the task of explaining how the system comes to be there to get inside of. The writer talks of his example as "the simplest form." Yet, in his own words, it is a *"powerful little chemical laboratory,"* well stocked with catalysers and other potent means for carrying on its work. "Simple"! Well, no doubt comparatively simple, but in reality complex almost beyond the power of words to describe. "A chemical laboratory"! Yes; and one which performs most delicate operations. "Well stocked with catalysers"! And what are they? Most wonderful things which induce change without themselves undergoing any; [Pg 171] discoveries of quite recent date as to which we still know but little. "Simple" seems hardly the word to apply, save in strict relation to other and higher forms. How did this laboratory come into existence? In what way did it learn to do its work? How did catalysers come to be? Was all this mere chance-medley? It is Paley's example of the watch found on the heath once more. Does it help us in any way to talk about "energy" and "complexes" of energy and "the creative force of energy"? To us it does not seem to advance matters one little bit. Either these operations of *Nitrosomonas* are determined or they are not; either they are the result of a law or they are the result of blind chance; in either case the energy which is involved must act according to the conditions ordered or not ordered. In other words: if it is the dominant factor, as the writer would lead us to suppose; if there is "direction," then the action of energy must be directive; and, if it is directive, in what possible way does it differ, save in name, from the old *entelechy* or *vital principle*, or whatever else one may choose to call it? On the other hand, if there is no such a thing as direction, if everything happens by chance, if the mechanistic theory is right, how does energy save us from complete surrender to that theory?

From all this it would appear that whilst energy is constantly being exhibited (and in all sorts of manifestations) by the living object, that does not explain anything, since it does not explain how energy originally came to be, nor how it came to [Pg 172] work under the laws which seem to govern it. It is one more added to the long list of "explanations," which hopelessly break down because those who have put them forward have never apparently applied themselves to the task of grasping the important difference between a final and an intermediate cause.

Let us sum up this part of our author's teaching in the light of this distinction. The organism is a material complex, and all sorts of actions and reactions take place in it. They are subject to the laws of physics, and notably to those relating to energy and its transformations. It has internal energies which must be adjusted to one another and not less to those around it; that is to say, it must be more or less in harmony with its environment. There are the problems of germ-plasm, and its transmission; the effect on it, if any, of the body, and the reaction of the body to its environment. There are also the catalysers of which we have spoken, with many problems associated with them, and throwing a possible and unexpected light on the vexed question of Vitalism and the Conservation of Energy. There are all these things, manifestations of energy; there is the watch, and it is going. But, as we remarked elsewhere, the fact that we have learned that the resiliency of the spring in the watch makes it "go" does not exhaust the explanation of the watch any more than the fact that we know something of the actions and reactions of energy in the organism exhausts its explanation. The watch is "going"; so is the organism. Each of them, [Pg 173] in a sense, is a "wonderful little laboratory" in which manifestations of energy are constantly taking place. The watchmaker constructed the watch for that purpose; who or what constructed the organism? Darwin and the Darwinians would have said—Natural Selection. In fact, Darwin rather lamented that "the old argument from design in nature, as given by Paley, which formerly seemed to me to be so conclusive, fails now that the law of Natural Selection has been discovered. We can no longer argue that, for instance, the beautiful hinge of a bivalve shell must have been made by an intelligent being, like the hinge of a door by man. There seems to be no more design in the variability of organic beings, and in the action of Natural Selection, than in the course which the wind blows." There again Darwin fell into a mistake, because he confused an intermediate with a final cause. Even if Natural Selection were all that the most ultra-Darwinian could claim it to be, it could not, as Driesch and others have shown, exhaust the explanation of the organism.

As a matter of fact the world of science is very far from thinking of Natural Selection as anything more than a factor, perhaps even a minor factor, in evolution. The author of the work with which we are dealing tells us that "Darwin's law of selection as a natural explanation of the origin of *all* fitness in form and function has lost its prestige at the present time, and all of Darwinism which now meets with universal acceptance is the *law of the survival of the fittest*, a limited [Pg 174] application of Darwin's great idea as expressed by Herbert Spencer." But let that pass. In another place the author makes it clear that the explanations of to-day, including his own, do *not* exhaust the subject, for he says "it is incumbent on us to discover the *cause* of the orderly origin of every character. The nature of such a law we cannot even dream of at present, for the causes of the majority of vertebrate adaptations remain wholly unknown." In any case we must account for Natural Selection; for if it is a Law—as some doubt—it must have had a Lawgiver. The watch must have been an Idea in some one's mind before it became an accomplished fact, and Natural Selection or any other "Law of Nature" must—unless all reason is nonsense and all nonsense reason—also have been an Idea before it became a factor. Whose Idea? Our author does not help us to answer this question. On the contrary—he tries to set an unclimbable fence in the way of any answer by telling us, though without any convincing argument to support his statement, that we may "exclude the possibility that it" [the internal moving principle] "acts either through supernatural or teleological interposition through an externally creative power." But though he refuses to allow us to look in this direction for a solution of our difficulties, it must be confessed that he does not help us with any other answer satisfying the question of the origin and evolution of Life.

FOOTNOTES:

[Pg 175]

[36] *The Origin and Evolution of Life; or, the Theory of Action, Reaction, and Interaction of Energy.* By F. H. Osborn. (G. Bell & Sons.)

[37] By *entelechy*—an Aristotelian term re-introduced by Driesch—is meant an agency other than one of a purely chemico-physical character, which differentiates living from not-living substance, and is responsible for the phenomenon of life.

INDEX OF NAMES

Agassiz, 142

Allen, Grant, 85

Aquinas, St. Thomas, 60, 147, 153

Austen, Miss, 32

Avicenna, 153

Balfour, Rt. Hon. A. J., 116

Bassi, Laura, 155

Bateson, W., F.R.S., 4, 7, 11, 118, 150

Bax, Belfort, 37

Benson, Mgr., 84, 88, 94, 101

Bergson, 151, 166

Bernhardi, 20

Borden, Sir Robert, 122

Branco, 162

Buffon, 100

Butler, Samuel 44, 61

Chesterton, G. K., 113

Clodd, E., 86

Conklyn, 23

Cowper, 37

Crichton-Browne, 20

Cuvier, 142

Darwin, 116, 131, 150, 173

Devas, Mr. 27, 120

Dewar, Prof. Sir J., F.R.S., 113

Doyle, Sir A. C., 46, 51

Driesch, 4, 7, 24, 69, 164, 166, 173

Fallopius, 96, 144

Fielding, 31

Gosse, E., 39

Gosse, Philip, 98

Grant Allen, 85

Healy, Father—Tale of, 40

Henderson, J. J., 167

Henslow, 24

Hull, Fr. E., S.J., 103

Huxley, 74, 98, 101, 117

Johnson, Dr. 48, 161, 168

Joly, Prof., F.R.S., 110
[Pg 176]

Kelvin, Lord, 151

Lankester, 15

Lauder, Harry, 2

Leduc, 2, 62

Lodge, Sir O., 3, 85

Loeb, J,. 58, 62

Lucas, E. V., on the War, 47

Mcdougal, 164

Mahaffy, Sir John, 111

Marett, 15, 16

Masefield, 48

Mendel, 75, 135

Milton, 145

Mivart, Prof., 96

Needham, John Turberville, 154

Newman, 33, 38

Newton, The Rev. J., 38

Nietzsche, 19

Osborne, Prof., 160

Paley, 160

Pasteur, 157

Perkin, Prof. W. H., 107

Pouchet, 157

Priestley, 156

Redi, Francisco, 153

Richardson, 31

Rignano, 25, 62

Ryder, Dr., 51

Sabatier, 113

Schwann, Theodor, 157

Scott, Prof., 142

Scott, The Rev. Thomas, 38

Sedgwick, Adam, 162

Spallanzani, Lazaro, 155

Stensen, Nicolaus, 75, 97, 99

Tilden, Sir William, 64

Tyson, Edward, 77

Wasmann, 26, 150

Wells, H. G., 49

Whiffen, 20

[Pg 177]

GENERAL INDEX

Adam, 146

Adrenals, 63

"After-Christians," 120

Aggressive mimicry, 123

Albino race, An, 128

Amazonian Indians, 20

"Anatomie of a Pygmie," 77

Ancestral peculiarities, 133

Aniline dyes, 107

Arrangement, 8, 137

Bacteria, Prototrophic, 169

Badische Aniline Fabrik, 106, 109

Bathybius, 98

Bion, 167

Blind Chance, 166

Bondage of Knowledge, The, 84

Botanic Garden, 131

Breeding Committees, 119

Breeding True, 126

Bricks and Builders, 139

"Bugbear of Hell," 21, 119

Calvinism, 32

Cartesian idea of the soul, 69

Catalysts, 113, 170

Celibacy, 120

Cell-Theory, The, 157

Chance-Medley, 134

Chromatin, 130

Colloids, 62

"Continuity," 46

Conversion, 34

Cowardice, Alleged, of Catholic Scientists, 99

Creation, 163;
a method of, 144

"Criticisms on the Pentateuch," 45

"Cutting up of Frogs," 115

Cytolysis, 65
[Pg 178]

"Dabney, Mr.," 47

Defence of the Realm Act, 82

Degradation of Energy, 168

Derivative Creation, 146

Discontinuity, 3

"Ducks and Drakes," 156

Duck's Egg, 125, 130

Dye-stuffs, 107

Elimination, Proof by, 163

Energy, 16

Energy, Degradation of, 169

Entelechy, 164, 171

Eskimo, 19

"Esmond," 31

"Essays and Reviews," 45

Eugenics, 117

Evangelicanism, 32, 33, 44

Exhibitions, International, of 1851 and 1862, 10

Extermination of the Less Fit, 122

Families, Restricted, 118

"Father and Son," 39

"Force and Energy, a Theory of Dynamics," 85

"Force of Truth, The," 38

Formaldehyde, 2

Fossils, Explanation of, 97

Free First Cause, 144, 151

Freethinkers and "tolerance, justice, and gentleness," 73

Germination, 65

Guide, the Church a, 92

Hapsburg lip, The, 127

Harmonious-Equipotential System, 69

Heredity in the Law Courts, 29

Hormones, 63

Horse, Pedigree of the, 161
[Pg 179]

Imprimatur, The, 77

In-and-in breeding, 127

Index Prohibitorius, 95

Industrial Scientific Research, Department of, 114

Inheritance:
Chemical theory, 134;
Mnemic theory, 5, 61, 133;
Particulate theories, 61, 132

Jack, Jill, and Joan, 119

Jungle, The law of, 122

King-crabs, 145

Lamp-shells, 145

Law and Heredity, The, 129

Law and Lawgiver, 9

Law of Nature, 174

Law's "Serious Call," 31

Liberty, personal, 87

"Life and Habit," 61

Life, Origin of, 160

"Little Dorrit," 112

"Loss and Gain," 33

Maggots in meat, 153

Man's pedigree, 161

"Marriage," 49

Mauve, 107

Mediate Creation, 147

Memory, unconscious, 5

Mendelism, 6

Method of Creation, 144, 161

Micromeristic theories, 5

Mimicry, 123

Mnemic Theory of Inheritance, The, 5, 61, 133

Monastic Orders, 121

Monophyletic evolution, 151

"Multitude and Solitude," 48

"Naturalism and Agnosticism," 57

Natural Selection, 19, 122, 173

"Nature does this," 136, 162

Nature's insurgent son, 15

"New Republic, The," 56

"New Revelation, The," 46, 51

Nitrobacter, 170

Novels and Novelists, 30

"Occam's" razor, 29

Occultism, 28, 51

Ordered energy, 166

"Organism as a whole," 38

Origin of Species, 150

"Over Bemertons," 47

Oxford Movement, 33

"Pamela," 32

Pangenesis, 61, 131

Pantheism, 9

"Paradise Lost," 145

"Parson Adams," 31

Particulate Theories of Inheritance, 61, 132

Personal Liberty, 81

"Philosophy of Biology, The," 163

Phylogeny, 4, 149

Plymouth Brethren, 99

Political leaders of the day, 114

Polyphyletic hypothesis, The, 150

Porto Santo rabbits, 148

Post-Christians, 27

Prototrophic bacteria, 169

Providentissimus Deus, 103

Pugs and Greyhounds, 126

Purposefulness: a strange confession as to, 59

"Raymond," 51

Resiliency, 172

Restricted families, 118

Sabbatarianism, 36

Salaries of Scientific Teachers, 112

Saurians, 145

Science, Catholic Men of, 75-6

Science, Neglect of, at Schools, 109

Sin, Mythical Ideas of, 123

Six-fingered race, A, 128

Slavery in the State, 24
[Pg 181]
"Slime of the Earth," 146

"Social Vermin," 118

"Some Revelations as to 'Raymond,'" 53

Special Creation, 142

Spermatozoon, 65

Spiritualism and the War, 50

Spontaneous Generation, 152

Springs in the watch, The, 172

"Stinks Men," 110

Survival of the Fittest, 122

Syngamy, 65

Synthetic drugs, 107

Telepathy, 2

Teratomata, 65

Theophobia, 26

Thermos flask, The, 113

"Throws back," 128

Trilobites, 145

Trinity College, Dublin, 110

"Tyranny" of the Church, 91

Uncle Remus and the rabbit's tail, 127

Unconscious Memory, 5, 61

Universities, Mediæval, 75

Vitalism and Anti-Vitalism, 68, 165

"Way of All Flesh, The," 44

"Wisdom, Book of," 123

Wolff's Experiment, 69

[Pg 182]

PRINTED BY
HAZELL, WATSON AND VINEY, LD.,
LONDON AND AYLESBURY.

www.ingramcontent.com/pod-product-compliance
Lightning Source LLC
Chambersburg PA
CBHW031426210526
45464CB00005B/2069